SMALL-SCALE MILLING

SMALL-SCALE MILLING

A guide for development workers

Lars-Ove Jonsson (Agrisystems)

David A. V. Dendy and Karen Wellings (NRI)

Varis Bokalders (SEI)

Intermediate Technology Publications
in association with
The Stockholm Environment Institute 1994

NOTICE

Neither the publisher nor the Agrisystems, the Natural Resources Institute, the Stockholm Environment Institute, the Swedish Mission Council and the Swedish International Development Authority make any warranty expressed or implied, or assume any legal liability or responsibility for the accuracy, completeness or usefulness of any information, apparatus, product or process disclosed, or represent that its use would not infringe privately owned rights. Reference herein to any specific commercial products, process or service by trade name, mark, manufacture, or otherwise, does not necessarily constitute or imply its endorsement, recommendations, or favouring by the authors, sponsors or publishers of this guide.

The guide includes information on the products and services of some manufacturers and suppliers of milling equipment. Inclusion of specific manufacturers does not constitute or imply endorsement, nor should omission of a manufacturer, supplier or product be considered significant in any respect.

The authors, sponsors and publisher assume no responsibility for any personal injury or property damage or other loss suffered in activities related to information presented in this book.

Published by Intermediate Technology Publications Ltd
103/105 Southampton Row, London WC1B 4HH, UK

© IT Publications and the Stockholm Environment Institute 1994

ISBN 0 85339 219 7

Typeset by Dorwyn Ltd, Rowlands Castle
Printed by SRP, Exeter

CONTENTS

FOREWORD   xi
PREFACE   xiii

1. INTRODUCTION   1

    Why Milling?   1
    Complexity of Choosing Appropriate Milling Techniques   1
    Need for a Manual in Small-scale Milling   2
    Approach   2
    The Target Reader   3
    The Content   3
    Definitions   3

2. THE CEREAL GRAIN   4

    Anatomy of Cereal Grains   4
    Nutrients in Cereals   5
    Purpose of Milling   5
    Phytic Acid   7
    Soaking   7
    Malting   7
    Fermentation   8
    Parboiling   8
    Type of Flour   9
    Recommendations on Flour Quality   10

**3. ELEMENTS OF TECHNICAL CHOICE
   IN MILLING**   12

    Case Study 1: TRADITIONAL MAIZE
    PROCESSING IN EASTERN PROVINCE
    OF ZAMBIA, AND MALAWI   12
    Background   14
    Traditional maize processing   14
    Comments   15
    Experience of sorghum and millet   15

    Case Study 2: MECHANIZED MILLING
    OF MAIZE IN EASTERN PROVINCE
    OF ZAMBIA, AND MALAWI   16
    Background   16
    Mechanized maize milling with hammermills   16
    Comments   16
    Experience of sorghum and millet   18

Elements of technological choice in milling   19
   ○ Milling enterprises   19
   ○ Homestead milling   19
   ○ Customs milling   19
   ○ Large-scale commercial mills   20
   ○ Consumers' preference   20
   ○ Retail prices   20
   ○ Geographical distribution   21
   ○ Utilization efficiency   21
   ○ Composite milling   21

4. PROCESSING REQUIREMENTS OF CEREALS, CASSAVA, LEGUMES AND CONDIMENTS   22

Wheat   22
   ○ Grain structure and composition   22
   ○ Wheat milling   23
   ○ Small-scale milling opportunities   24
Maize   24
   ○ Grain structure and composition   24
   ○ Milling considerations   24
   ○ Small-scale milling opportunities   24
Rice   25
   ○ Grain structure and composition   25
   ○ Milling considerations   26
   ○ Small-scale milling opportunities   26
Sorghum   27
   Grain structure and composition   27
   Small-scale milling opportunities   27
Millet   29
   ○ Grain structure and composition   29
   ○ Milling considerations   30
Barley and Oats   30
Cassava   30
   ○ Structure and composition   30
   ○ Processing considerations   31
Legumes   32
   ○ Structure and composition   32
   ○ Processing considerations   32
Condiments   32
   ○ Roots   32
   ○ Seeds   32
   ○ Bark   32
   ○ Dried whole fruits   32
   ○ Dried bulbs   32
   ○ Dried leaves   33
Salt   33
Bone Meal   33

5. MILLING PROCESSES   34

Pre-treatments   34
- ○ Grain cleaning   34
- ○ Foreign material   34
- ○ Pesticide residues   35
- ○ Peeling and drying roots   35
- ○ Legume splitting   35
- ○ Moisture and drying   35
- ○ Cleaning   35

Husk and Bran Removal   36
- ○ Manual husk and bran removal   36
  - ● Pestle and mortar   36
  - ● *Dheki*   36
- ○ Mechanized husk and bran removal   36
  - ● Abrasive disc dehuller   36
  - ● Steel huller   37
  - ● Centrifugal dehusker   38
  - ● Rubber roll dehusker   39
  - ● Under-run disc dehusker   40
  - ● Friction whitener   40
  - ● Cone whitener   42
  - ● Abrasive whitener   42

Grinding   43
- ○ Manually operated grinders   43
  - ● Pestle and mortar   44
  - ● Saddlestone   44
  - ● Quern   44
  - ● Hand-operated rotary mill   44
- ○ Mechanized mills   45
  - ● Stone mills   46
  - ● Plate mills   47
  - ● Hammermills   48
  - ● Roller milling   51

6. SOURCES OF POWER   52

Human Power and Draught Power   52
Diesel Engines   54
- ○ Old-fashioned diesel engines   55
- ○ The modern diesel engine   55
- ○ The hot-bulb engine technique   56
Electric Motor   57
Water-mills   57
Wind Power   58
Battery-powered Mills   58
Steam Engines   60
Coupling Arrangements   60
- ○ Direct drive   60
- ○ Flat belt drive   61
- ○ V-belt drive   61
- ○ Centrifugal clutch coupling   61
- ○ Multiple drives   62

7. OPERATOR'S MANUAL FOR DIESEL-POWERED HAMMERMILL   64

Introduction (Case 3.1)   64
The Hammermill Operation   64
  ○ Daily pre-operation tasks   64
    ● Checking and filling of oil (Case 3.2)   65
    ● Checking, filling and recording of fuel (Case 3.3)   66
  ○ Engine operation tasks   66
    ● Starting and stopping of engine (Case 3.4)   66
  ○ Milling tasks   67
    ● Milling and recording grain (Case 3.5)   67
  Post-operation tasks
    ● Cleaning hammermill and shelter daily (Case 3.6)   68
Hammermill Maintenance   69
  ○ Weekly maintenance tasks
    ● Check and tighten selected bolts (Case 3.7)   69
    ● Check and tighten beater bolts (Case 3.8)   69
    ● Cleaning of air filter elements (Case 3.9)   70
    ● Cleaning shelter and divider (Case 3.10)   71
  ○ Monthly maintenance tasks
    ● Cleaning of cooling fins (Case 3.11)   71
    ● Checking and adjusting V-belts (Case 3.12)   72
  ○ Maintenance tasks after 300hrs of operation
    ● Greasing of mill bearings (Case 3.13)   73
    ● Changing of oil and replacing oil filter (Case 3.14)   73
    ● Changing of fuel filter element and cleaning of ceramic
      filter (Case 3.15)   74
    ● Changing of air filter element (dry type) (Case 3.16)   74
    ● Removing soot from exhaust silencer (Case 3.17)   75
Hammermill Repairs
  ○ Removing and replacing of damaged screen (Case 3.18)   75
  ○ Turning or replacing beaters/hammers (Case 3.19)   76
  ○ Changing of damaged V-belts (Case 3.20)   76
Major repairs
  ○ When and how to get outside repair service (Case 3.21)   77
Management
  ○ Record keeping (Case 3.22)   77
  ○ Daily Repair Form (Case 3.23)   78
  ○ The Weekly Records Form (Case 3.24)   78

8. INSTALLATION AND SAFETY OF MILLING   81

Installation and Layout of Equipment   81
  ○ Basic requirements   81
  ○ Hand-operated mills   81
  ○ Animal-powered mills   81

Motor-powered Mills   81
    ○ Diesel engine   81
    ○ Electric motor   83
    ○ Steam engine   83
Water-powered Mills   83
Wind-powered Mills   83
Safety   83

9. SUMMARY OF MILLING PROBLEMS   85

10. SMALL-SCALE MILLING MACHINERY   87

11. MANAGEMENT   89

Introduction   89
How to Choose a Mill   89
Comments   90
Financial Costing   91
    ○ Pay-back method   91
    ○ Budgeting procedure   91
Follow-up   93

Case Study 4: MANAGEMENT   94
    ○ Daily Records Form   95
    ○ Weekly Records Form   96
    ○ Cash Book   97
    ○ Milling Project Description   98
    ○ Profitability Projection   101
    ○ Cash-flow Analysis   104

APPENDIX 1: Glossary   106

APPENDIX 2: Further Reading and Information   108
References   108

APPENDIX 3: List of Useful Contacts   110

APPENDIX 4: Product Details   114
Machinery Suppliers and Manufacturers   114
Further List of Manufacturers and Suppliers   125

APPENDIX 5: Sources of Illustrations   128

FOREWORD

Milling is an essential task for millions of women in developing countries who work with pounding or milling grains to make the daily food. Manual milling by primitive means is very time-consuming, which in many cultures meant that local mechanized mills have been established. Depending on the energy sources available, they could be watermills, windmills or animal-powered mills. Later on in the development process bigger mills were often established where milling was done in a more centralized way. But in many developing countries small local mills still have a very important role to play.

Poor people in rural areas are often excluded from milling in big centralized mills. Partly because this is too expensive and partly because transport is difficult to get. The fieldworkers of the Swedish Mission Council realized that this was a big problem and wanted to do something about it. So they asked the Stockholm Environment Institute to co-operate with them in producing a guide for development workers on small-scale milling.

As we did not have the expertise in this field we looked around for partners who were knowledgeable in the field of milling. We thus asked our colleagues at the Natural Resources Institute in England and at Agrisystems in Sweden to help us write this book.

Mr Varis Bokalders was appointed Project Leader and he formed a team of experts to look into this issue. The team consisted of himself, Mr Lars-Ove Jonsson, Agrisystems (Nordic) AB, Uppsala, Sweden and Dr David A. V. Dendy and Mrs Karen Wellings of the Natural Resources Institute in the UK. The team members of NRI were assisted by G. Anstee, J.S. Bickersteth, P.A. Clark, J. Coulter, S. Harding, A.W. James and N.H. Poulter, all from NRI.

Of the 11 chapters, Mr Lars-Ove Jonsson wrote chapters 1, 2, 3, 7, most of chapter 11 and parts of chapters 6, 8 and 9. In addition he developed the outline, edited the text, scanned and edited the figures and finally performed the desk-top editing except for the illustrated list of manufacturers. NRI members did the basic research and wrote the remaining parts.

The book looks into the milling process, the main different cereals, and into the small-scale technologies available for milling. It also describes some case studies, in order to make the reader more familiar with all aspects of local milling, like economy, management and operation.

We hope that this book will make it easier for people to establish small local mills, and that this will help in the development process by making life easier for women in the rural areas in developing countries.

Lars Kristoferson
Stockholm Environment Institute
November 1993

PREFACE

Small-scale Milling: A guide for development workers is a collaborative publication involving the Stockholm Environment Institute, Agrisystems and the Natural Resources Institute. The book is published by IT Publications Ltd and is the fourth in a series, the first being *Solar Photovoltaic Products*: the second being *Micro-Hydro Power* and the third being *Windpumps*.

The book is essential reading for anyone involved in the milling of cereals in developing countries and the target reader is the extension agent promoting improved small-scale milling. It covers:

- technical aspects of milling;
- economic aspects of milling;
- social aspects of milling;
- nutritional/health aspects of milling.

ACKNOWLEDGEMENTS

The authors wish to thank the Swedish International Development Authority, SIDA, for supporting the production of this guide and providing the finance for the research work. We also gratefully acknowledge the information and advice provided by the local and international organizations involved in promoting improved small-scale milling and the international milling equipment industry. Particularly, the information on fermentation from the Department of Food Science, Chalmers University of Technology has influenced the content of the book.

For further information, please contact:

Varis Bokalders
The Stockholm Environment Institute
Box 2142
S-103 14 Stockholm
Sweden
Telephone: 46-8 723 02 60
Fax: 46-8 723 03 48

Karl-Erik Lundgren
Swedish Mission Council
Office For International
    Development Co-operation
Tegnergatan 34 n.b
S-113 59 Stockholm
Sweden
Telephone: 46-8 30 60 50
Fax: 46-8 31 58 28

Lars-Ove Jonsson
Agrisystems (Nordic) AB
Sandmovägen 6
S-756 47 Uppsala
Sweden
Telephone: 46-18 30 18 96
Fax: 46-18 30 06 18

David A. V. Dendy and Karen Wellings
Natural Resources Institute
Central Avenue
Chatham Maritime
Kent
ME4 4TB
United Kingdom
Telephone: 44-634 88 00 88
Fax: 44-634 88 00 66/77

INTRODUCTION

Why Milling?

Cereals have been staple food for thousands of years, since people first began to settle in permanent communities. Wild cereal grasses were the early food grains and they had to be processed before consumption. The old practice was to crush the seed into smaller pieces which broke off the indigestible, fibrous outer parts from the desired nutritious inner parts. The latter was further ground into a coarse flour, which was cooked to a porridge or baked as a flat bread. Even in the early days, grains were not food until they had been processed (crushed and cooked). The grain was crushed or milled between stones or in wooden mortars, with pestles. The earliest milling technique used is believed to be the saddlestone which dates back to 10000-6000 B.C. in Mesopotamia and Egypt and was often pictured in Egyptian wall paintings. The pestle and mortar is an equally old milling technique.

The saddle-stone was equipped with a handle to move the stone forwards and backwards to improve efficiency and by approximately 500 B.C. the rotation motion of the stone was introduced. Greeks and Romans used rotating stone mills (hour-glass mills) from 400 B.C. These mills have been found in Pompeii and Ostia. Milling of grain was a serious constraint for the Roman empire. How to feed the city of Rome and the large army with wheat flour? Donkeys and oxen were used to drive the mill which increased the milling capacity substantially. The rotating milling technique spread throughout the Roman Empire and a system was developed to adjust the distance between the stones to allow for different fineness of flour. The stones were also dressed in different patterns to improve the rate of grinding.

The first water-mills mentioned in text date from about 100 B.C. (Antipater, Strabo, Vitrivius) and they had either horizontal or vertical milling stones. Windmills are first reported in Persia 600 A.D. and came to England, France and Flanders about 1000-1100. Windmills were spread all over Europe with the monasteries. The modern milling technique used today was developed during the last century to feed the fast-growing urban population and the emerging livestock industry.

Despite the introduction of foreign grain varieties, new food products and food habits, milling is still a major operation in the food preparation chains of cereals. The old milling technique is still dominant in many rural societies of the developing world and one can say that few techniques have been more resistant to change. Normally, there is a reason for a change of technique and the maintenance of the old technique over such a long period of time in many rural areas is a strong indication that the operation is essential for the survival of man and that the technique applied is the best for the local food processing conditions.

There is a common difference in the perception of the importance of the milling process in the food production chain between people raised in the developed world and the rural poor in developing countries. In many cases it is fair to include the urban elite from developing countries in the former category. When people in a developed country need flour, they simply buy it from a shop and few people would recognize the efforts involved in a packet of enriched flour from the grain in the farmer's field to the shop as it is produced and delivered by some few specialist industries. The situation is very different for the many rural women in developing countries as they often have to undertake both the job of growing the grain, the process of milling and the preparation of porridge or bread to feed their families by using old hand-tool techniques. This dilemma of different understanding of problems is often reflected in aid programmes which tend to give priority to activities which increase production while many of the recipient households would prefer additional assistance with other tasks like grain processing.

Complexity of Choosing Appropriate Milling Techniques

Traditionally, women store and prepare the grain into food for the family, and milling is an integrated component of the cereal food preparation chain. The milling techniques are often specific to a particular region or culture, however, certain basic principles are applied universally.

The female members of the rural households will devote 2-4 hours of hard labour per day in preparing the daily requirement of flour for their family's needs in many developing countries. Mechanizing the milling is a desirable step in reducing time and effort spent on milling. However, any changes in the old milling techniques will have implications, both positive and negative, in relation to:

- gender roles
- food preferences
- nutrition and health
- food security
- technology/productivity
- socio-economic factors
- income and/or costs
- sustainability

Men will normally take over milling, once it has become mechanized. This is particularly true with customs milling involving money. Mechanized milling will reduce the workload of women, but the new technique can also have serious implications for the nutrition of children and pregnant women, who are dependent on a cereal diet. Many milling projects produce a type of flour determined by technical efficiency and not by the customers' desire or need. 'Improved milling technique' means normally scrapping the traditional technique of superior nutritional value. Milling charges increase the need for cash which tends to increase the workload of women who are usually the main cash earners. She might win on the one hand and lose on the other. Another important aspect is that mechanized mills break down and can be out of order for longer periods. What alternative do women have when there is no customs mill working in the vicinity?

The mechanization of rural, small-scale milling processes will most likely lead to conflicts of interest from the farm to the national level. There is a tendency to underestimate the complexity by dealing only with the technical aspects of the mechanization of milling. A narrow technical approach runs the risk of leaving a number of bottle-necks and issues unattended, which limits the opportunities to promote sustainable rural processing systems. When designing and managing appropriate milling techniques, it is important to work with a broad analysis incorporating the above-listed aspects. Technology should aim at the best combination of positive factors, not only for millers and traders but also for the consumers of milled products.

Need for a Manual in Small-scale Milling

A number of NGOs, especially those working with African problems, have experienced difficulties in improving traditional milling both from technical and nutritional points of view. They particularly lack practical guide-lines on how to design and manage appropriate new milling techniques in rural areas.

The Stockholm Environment Institute, upon request from the Swedish Mission Council, has decided to form a group of experts to prepare guidelines on small-scale milling with specific reference to the African situation. However, due to many similarities, the guide-lines can be used in other countries in Asia and Latin America.

Approach

The work of grinding cereals to flour for human and livestock consumption, on farm, homestead or at village level, is an integrated part of a farming system where subsistence production prevails. Small-scale milling in urban areas is, on the other hand, a small-scale industrial activity servicing urban dwellers. The common trend among many development agents in Africa has been to look at rural milling as a non-agricultural activity. The industrial approach has a tendency to disregard many of the specific rural constraints as well as advantages, resulting in development of sub-optimal processing systems. To some extent this means the development of mechanized small-scale milling systems with built-in shortcomings which could have been minimized if they had been derived more closely from the existing farming systems.

The intention of this book is to view the small-scale milling problems more from a farming systems point of view. In addition to the common industrial tradition, this will include more considerations of the usefulness of the milling techniques and the milled products for the rural households. No doubt, this concept will also be beneficial for milling projects servicing poor city dwellers.

An example of the above mentioned is local fermentation of the grain before milling. This pre-milling activity is mostly neglected by the industrial approach in east and southern Africa while it is considered vital from the farming systems view as it improves the utilization of nutrients of the milled products.

The Target Reader

Improved small-scale milling techniques in rural areas will have to cover:

- technical aspects;
- economical aspects;
- social aspects;
- nutritional/health aspects.

An important part in appropriate small-scale milling promotion is the development of knowledge and skill to deal with the above-listed aspects in a more efficient way.

Rural entrepreneurship is a delicate undertaking in most developing countries with economic problems such as inflation rates of more than 100 per cent/year. Constant and fast adjustment changes of prices and markets are a must if the business is to remain in operation. A sustainable business enterprise can not only cope with the changes but can also maintain or increase its market share of desired products. This can be a difficult problem when the milling charges increase more than the household income.

Promoters of improved small-scale milling techniques (technical extension agents) will play a key role in encouraging appropriate rural entrepreneurship and they must possess the knowledge and skill to design and manage sustainable milling operations. Unfortunately, there are few people trained to perform this task to the required standard. An obvious conclusion is therefore, that the target reader must be the technical extension agents (TEAs), home economics agents and nutritionists dealing with small-scale milling either from the agricultural or industrial side — this embodies the 'train the trainer' concept.

The Content

The content of the book will cover necessary components to enable the target readers particularly to develop the ability to design and manage appropriate and sustainable small-scale milling enter-prises in rural areas. It is not possible to cover all conditions but the main theme is a farming systems approach for appropriate solutions both for the millers and their customers, especially the poor who depend mainly on a cereal diet as their staple food. The authors are convinced that this goal will be best achieved by presenting some basic theory of milling techniques, the nutritional value of milled products and management in combination with selected case studies. A TEA needs to know more than a mill operator and the book is designed to allow for more selective reading of parts only to satisfy the demand for specific groups of readers.

The book starts with a description of the cereal grain and its usefulness as food. Then come two case studies, one on traditional pounding and the other on an early attempt to mechanize maize milling followed by a description on milling enterprises. Chapters 4-6 deal more specifically with the processing requirements and a detailed study of milling techniques and equipment. Chapter 7 is a case study of an Operator's Manual for a hammer-mill succeeded by recommendations on installation and safety of milling and a check-list of milling problems. The next part covers basic data and the book ends with a chapter on management combined with a case study of this subject.

Definitions

Due to differences in the understanding of the basic terminology used in milling, there is a need to define some of the basic expressions. It is also important for the readers to know the meaning of the below- mentioned keywords in milling.

Milling is a trade term meaning the reduction of grain to meal or flour. Milling as an overall process, includes size reduction, dehusking/ dehulling, scarifying, polishing, sorting, mixing and also chemical reactions like fermentation. In this book, to avoid misunderstanding we will use the term grinding when the grain is crushed to meal or flour.

The term dehusking is used for the operation to remove the outer fibrous husk (fused glumes, see Figure 1) from rice, barley, etc. Dehulling, on the other hand is used for removing the underlying bran and aleurone layers (pericarp and seed coat, see Figure 1).

THE CEREAL GRAIN

2

Anatomy of Cereal Grains

All cereal grains, wheat, maize, sorghum, rice, etc. belong to the same grass family (Graminae). Although the grain kernels look different, the structure of each grain is basically the same. For milling purposes, however, it is useful to make a distinction between naked and husked grain. Wheat, maize, sorghum, etc. belong to the naked group while rice, barley, oats and some types of millet belong to the husked category.

The basic grain structure is shown in Figure 1.

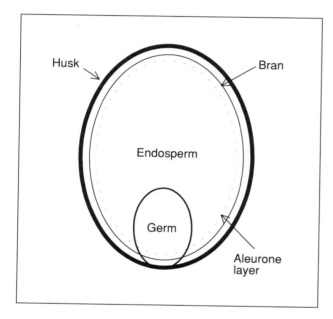

Fig. 1 *General structure of cereal grain*

A cereal grain consists of five main components:

- **the germ**, or embryo from which a new plant develops, rich in protein, fat, vitamins (vitamin E) and minerals;
- **the endosperm**, the starchy interior;
- **the aleurone** layer, a thin layer rich in protein, fat, vitamins (vitamin B) and phytic acid;
- **the bran**, a rich source of protein and fibre;
- **the husk**, the external fibrous part.

The aleurone layer forms the outer part of the endosperm and the milling process used will determine if it will be a part of the endosperm or the bran. In practical milling terms, it is feasible to talk of only four components: the germ, the endosperm, the bran and the husk.

The term husk is used to describe the totally indigestible part, which is the outermost layer to be seen in rice, oats, barley and some millet varieties. Other grains are threshed from the stalk without husk, but have bran layers which are often difficult to digest and these are known as hulls. Legumes are similar, but more rich in protein. Condiments (spices, herbs, colourings) range from dried roots to numerous species of seeds and dried leaves or stalks.

Figure 2 shows the approximate proportion of the cereal grains. The millet variety shown is pearl millet. More detailed information on four of the grains can be found in Table 1 opposite.

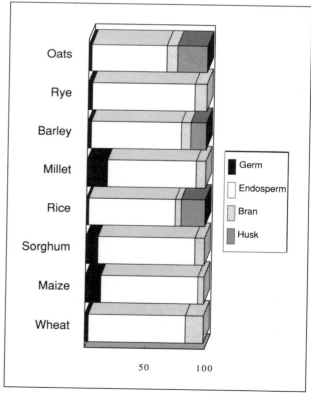

Fig. 2 *Proportion of parts in cereal grains (%)*

4

Depending on the structure of the grain used and the final use of the milled product, the various parts of the grain may be separated to a greater or lesser extent during the milling process. This is true also of dried roots such as cassava, where milling gives a flour with some coarse fibres.

Nutrients in Cereals

Table 1 shows the approximate composition of nutrients in four major types of grain. There are large variations due to different varieties and cropping conditions. The data are averages from a number of sources. The soluble carbohydrates consist mainly of starch granules and the ash column indicates the mineral content in various parts of the kernel.

The distribution of nutrients within the cereal grains shows a number of general features and some significant individual differences between various cereals. Starch is generally present only in the endosperm, but protein occurs throughout the grain with a concentration in the germ and the aleurone layer. Fat (oil) is concentrated in the germ, the aleurone layer and the bran. Conversely, the endosperm and germ are relatively free from fibre which is found in the husk, bran and aleurone layers.

Cereals are an important source of protein but the cereal protein is deficient in a vital amino acid: lysine. Maize and sorghum is also low in tryptophan. The biological value of the protein in the germ and aleurone layer expressed as lysine content is 2-2.5 times higher than that of the endosperm protein.

The cereal fat (oil) from the germ is of good quality, but the germ must be separated from the grain before oil can be extracted for commercial use. Fat-soluble vitamins like vitamin A and E are mainly found in the germ.

The ash content shown in Table 1 indicates that the mineral content is low in the endosperm but high in the germ and the bran. There is a heavy concentration of vitamin B in the aleurone layer and the germ compared to the endosperm. In general, cereals are low in vitamin A and C.

Table 1. Approximate composition of cereal grains (%, dry matter basis)								
PRODUCTS	CEREAL	KERNEL %	SOLUBLE CARBOHY-DRATES %	PROTEIN %	FAT %	SUGAR %	FIBRES %	ASH %
WHOLE GRAIN	WHEAT	-	79.8	12.6	2.6	2.5	2.5	1.8
	MAIZE	-	80.3	10.8	4.8	2	2.1	1.7
	SORGHUM	-	78.5	12.4	3.6	1.1	2.7	1.7
	RICE	-	71.2	9.1	2.2	-	10.2	7.2
ENDOSPERM	WHEAT	82	85	11.4	1.2	0.8	0.1	0.44
	MAIZE	82	86.4	9.4	0.8	0.6	0.1	0.3
	SORGHUM	82	86.4	9.5	1	0.4	1	0.8
	RICE	73	88.9	9.8	0.5	-	0.3	0.6
GERM	WHEAT	3	20	29.4	10	20	8.9	8.9
	MAIZE	13	19	18.8	34.5	10.8	4.6	10.1
	SORGHUM	10	21	15.1	20	12	2.6	8.2
	RICE	2	-	-	-	-	-	-
BRAN+ ALEURONE LAYER	WHEAT	15	1	11.1	3.5	0.3	13.5	6.1
	MAIZE	5	7	7.1	1	0.3	14	5.8
	SORGHUM	8	5	8.9	5.5	0.3	8.6	2.4
	RICE	5	10	14.1	18.8	0.5	5.7	11.2
HUSK	RICE	20	34	1.2	0.3	0.3	34	24
Source: Summary of various sources (average figures)								

Purpose of Milling

The human digestive system has limitations in utilizing the whole grain efficiently even if it is boiled or roasted. The efficiency will improve with the reduction of grain size to smaller particles to allow for larger surface for the digestive fluid to act on.

The main objectives of milling cereals are to:

- improve the digestibility and other nutritional features;
- increase the variety of foods;
- allow for separation of husks, bran, germ and endosperm;
- reduce the fibre content.

This is achieved through breaking down the grain into smaller particles by mechanical means to free starch, protein, fat and fibres from the cell structure (see Chapters 3–5). The fractions may be separated to obtain desired products by various techniques including winnowing, sifting and flotation. Products may include husks, bran, germs, grits, dehulled/parboiled/polished grain, meal, flour, starches and protein fractions as well as grain oil. Sophisticated milling methods have been developed to produce these products, but in this book we will only discuss small-scale processing techniques appropriate for rural areas and limited to a number of products.

Cereal flour may be converted into leavened or unleavened bread, thick or thin porridge, pasta and other steamed, boiled, fried or roasted products. Whole grain or milled products may be fermented before or after germination. Beer may be brewed and consumed as opaque beverage before fermentation is completed. Whole or partly processed grain can be boiled or roasted.

Though causing a loss of some fibre, vitamins and fat, removal of the outer layers (bran, hulls) of the grain is usually desirable. The germ may be removed at this stage so that the final flour contains only a limited proportion of fats which may deteriorate by exposure to air and/or water. The risk of rancidity is thus reduced and storage stability much improved for marketing purposes.

Although some fibre is needed to absorb waste into the faeces and to minimize lower gut problems, a high fibre content in the diet will on the other hand, reduce the efficiency of digestion of proteins, fats and starches and lower the absorption of vital vitamins like A and E. Deficiency of fibre is hardly a problem of the rural population in developing countries but more of an urban disorder of the industrialized world.

In addition, high fibre content will also increase bulkiness that will lower the concentration of nutrients in the food. This can cause serious malnutrition problems when the food diet consists of mainly cereal products. Children with their small stomachs are particularly prone to suffer as they get full before having satisfied their nutritional needs. Fibre is also reported to reduce the capacity of the bowel to extract minerals from the cereal food.

The aleurone layer near the surface of the grain contains a substance, phytic acid (see section on phytic acid), which interferes with the absorption of minerals (forming insoluble compounds).

The proportion of the whole grain that becomes flour or meal during the milling process is known as the extraction rate. Low extraction flours and meals are usually whiter and of lower fibre content than whole — 100 per cent extraction — flours, and are better digested due to less interference from phytic acid. Removal of other anti-nutritional factors, such as the pigment tannin which is concentrated in the outer layer of sorghum, is also nutritionally desirable. Tannin binds both minerals and proteins, preventing absorption into the body. In addition, tannin causes digestive problems like constipation.

Dehulling (removing of bran) is a common practice to reduce the fibre, phytic acid (50-60 per cent reduction) and tannin before grinding the grain into flour.

The significance of nutritional changes associated with milling for health have to be considered against the background of the diet. At the village level where food supply and variety is limited, and where emphasis is on locally available crops, it is important to add other foods to the diet, especially when cereals form the main part. Unfortunately, many people are not aware of the nutritional side-effects caused by changed milling practices and/or many cannot afford the supplementary food needed for a balanced diet.

Pellagra is caused by lack of a B vitamin (niacin) among people dependent on white maize flour (low extraction rate). Beriberi is caused by the deficiency of another B vitamin (thiamine) in dehulled rice. Vitamin E found in the germ helps preventing eye disorders in premature infants and elderly people and also anaemia and heart problems. No doubt, promoters of changed milling practices have a great responsibility for many people's health and particularly children's well-being.

Cassava is a starchy root which, correctly processed, becomes high energy food. It has little protein, fat, or vitamins apart from vitamin C and it does contain a toxin (poison), so the methods of processing cassava to remove this are described.

Legumes — beans, peas, lentils and other pulses — are briefly described. They are a rich source of protein, though some may contain harmful substances that require washing out or destruction by heat.

While condiments may not be particularly nutritious — though some are — they are essential for making food more acceptable. Many condiments contain healthy substances that help the body resist sickness: a very brief account is given of their milling requirements.

Phytic Acid

Phytic acid which is found in cereal crops, has phosphorus (P) groups attached to it which form complex bonds (insoluble bindings) with essential minerals like iron, calcium, magnesium, sodium and zinc from the diet, whether eaten in the cereal food or in other items of the diet. Recent findings also indicate that phytic acid affects digestibility of protein and starch. This particularly explains the high prevalence of iron-deficiency anaemia in Africa, making it the most serious nutritional problem after malnutrition. Phytic acid is also a main contributor to rickets (the 'English disease') which deforms the body of growing children due to vitamin D and calcium deficiency in areas which depend on non-fermented whole-meal products as their main food. Phytic acid is an anti-oxidant which increases the stability of minerals and other compounds in cereal products.

By activating the enzyme phytase which is available in the seed, the phytic acid will be degraded and the bioavailability of minerals will increase substantially. The natural methods to activate phytase are soaking in water, germination and fermentation. (hydro-thermal processing).

The naturally occurring phytase in cereals is activated by soaking under optimum conditions (pH 4.5–5.5). Soaking of wheat bran will then result in a 95 per cent hydrolysis of phytic acid within one hour and a complete degradation within two hours. This will increase the bioavailability of iron by a factor of six. The optimal effect of phytase is at a temperature around 55° Celsius.

Malting is a process during which the whole grain is soaked and then germinated. During this process a number of enzymes including phytase are activated provided the pH is kept within 4.5–5.5. At this pH level, ground malted cereals will have almost all phytic acid degraded

Fermentation is an old practice for food processing and preservation. Due to the production of lactic acid and other organic acids, the pH is lowered and the phytase activated.

Hydro-thermal processing techniques and artificial phytase production are attracting great interest from the feed industry research as it improves digestion of feed, reduces the need of extra minerals in the feed and reduces the phosphorus content in manure which is an important water pollution issue in Europe. It is reported that pigs are able to utilize only 30 per cent of the available phosphorus in untreated cereals compared to more than 60 per cent when treated. In other words, the pig farmers do not need to apply additional minerals when cereals are fermented before feeding which is a necessity with untreated feed.

Soaking

Soaking is the initial step of malting, fermentation and parboiling (rice). The common time of soaking grain or grits is 24–48 hours in lukewarm water in traditional food processing. Soaking of grain in water for 24 hours before milling is enough to increase the amount of soluble iron up to 10 times. During the time of soaking, the lactic acid fermentation is developing and the pH drops. The germination process starts when the moisture content in the seed is raised to 42–46 per cent.

Malting

A pre-condition for germination is that the soaked grain is exposed to oxygen. After soaking the surplus water is drained off and the grain is spread out in a thin layer for proper airing while germination takes place. Frequent re-wetting and turning is needed to prevent drying-out and rotting. After two days the grain will germinate and when the sprouts are about 1 cm long, the grain is spread out in the sun to dry. When thoroughly dry, the shoots and root may be rubbed off and removed by winnowing. The germinated grain is now ready for grinding and use.

Germination, the start of the growth of the seed to a plant, releases other enzymes from the germ in addition to phytase. The germinated grain is called malt and is rich in enzymes breaking down the starchy, white part of the grain to sugar. If yeast is present, either by addition or native to the grain, then water is added to the ground malt and the sugar will break down to alcohol. This is the principle of beer making and both opaque beer and lager can be made from malt of most cereals.

Ground malt which contains the enzyme amylase is also used in porridge to increase nutritional value of baby food by increasing the actual nutrition per spoonful and to make desirable and healthy drinks like Milo or Bournvita.

Fermentation

Fermentation of grain before final grinding to flour is a traditional practice to improve digestibility and nutritional value of grain products, improve taste and texture and to extend shelf life of flour. Despite its many advantages, the development policies on mechanized small-scale milling has aimed at eliminating this step as outmoded in the food processing chain in many parts of eastern and southern Africa. In west Africa where wet milling is practised, fermentation is an integrated part of mechanized small-scale milling, but the flour produced this way can only be stored for a day before being spoilt.

Recent findings have shown that fermenting is a vital component both in weaning foods for children up to three years based on cereals and also for the nutrition of adults. The improvements are of such a magnitude to imply that fermenting ought to be included in any milling project aimed towards low-income groups both in rural and urban areas. The process will have to be adopted according to food preferences and the use of flour.

The changes occurring during the fermentation process are mainly due to enzymatic activity brought about by the micro-organisms and/or the indigenous enzymes in the grain. The micro-organisms fall into three categories:

* **Yeast** (bread, beer, alcohol);
* **Moulds** (cheese, legumes);
* **Bacteria** (cereals and animal products).

The two major types of bacteria in cereal and tuber fermentation are lactic acid- and acetic acid-producing bacteria. The micro-organisms involved in natural fermentation of cereals are essentially the surface flora of the seeds. During fermentation the pH drops from about 6.5 to 3.6. This effectively inhibits the growth of other bacteria that cause decomposition and food spoilage. They will also have a strong inhibitional effect on diarrhoea-causing pathogens which substantially reduces the risk for diarrhoea among children; 'the biggest child killing disease'.

Whole-grain flour made from fermented and dry grain can be stored for longer periods (increased shelf-life) compared to un-fermented flour. Particularly, the essential vitamin E will be maintained in the flour for a much longer period. Another noticeable effect of fermentation is a softer grain texture reducing substantially the energy required to separate fibrous parts and produce flour.

Lactic bacteria are reported to contain proteolytic activity capable of degrading more complex proteins into simple proteins, peptides and amino acids. This might explain the favourable effect on protein digestibility found after lactic fermentation of wheat, maize, sorghum and millet.

Fermentation of cereals inhibits the negative effects of phytic acid and tannin. It is evident that fermentation is an effective and low-cost method of preventing/reducing diseases caused by deficiency of important nutrients in regions where cereals form a major part of the diet.

The destruction rate of phytic acid appears stronger with bacterial fermentation compared to yeast fermentation only. This means that a sour dough will produce a bread made out of whole-meal flour with lower phytic acid content compared to a more neutral dough. Unleavened bread or bread made from baking powder will have the highest level of phytic acid. Porridge made from fermented cereals will subsequently have a higher nutritional value compared to being made from unfermented flour.

Parboiling

This ancient process is used mainly for rice, but also for wheat. The rice is pre-treated by soaking in water, heating to gelatinize its starch followed by cooling and drying. Conventional rice milling follows (see Chapter 5). Various potential benefits result from parboiling: grains are easier to dehusk and resist milling breakage; the nutritional profile of the grain is improved through better vitamin retention; and it has improved resistance to storage infestions. However, the resulting parboiled rice takes longer to cook and if the process is not properly carried out spoiled odours can develop.

Type of Flour

It is important to note that commercial mills in industrial countries adhere to established flour standards which include quality specifications and enrichment with iron and vitamin B and other additives to increase quality to meet customers' demand and to reduce the risks for deficiency diseases. The situation is different in most developing countries where enrichment with vital minerals and vitamins is rare. By integrating traditional practices like soaking, germination and fermentation in the milling process, it is possible to upgrade the common cereal diet from being of relatively low bioavailability standard of vital nutrients to intermediate/high bioavailability level. This means that the milling technique will have serious implications on the nutritional well-being of particularly poor people in rural and urban areas. Therefore, it is helpful to define a simple milling practice standard for small-scale milling when discussing the quality of the milled products in this book.

For practical purposes one can define three main types of qualities of flour relevant for small-scale milling in rural areas:

Qa - Whole-meal flour (97–99 per cent extraction rate);

Qb - Dehulled and/or partially sifted and degermed meal (80–90 per cent extraction rate);

Qc - Degermed, dehulled and sifted flour (60–75 per cent extraction rate).

Different fineness of grain particles is required for different products. Taste and preferences differ but one can specify according to common use:

- flour (f) – cake flour, flat bread, etc.;
- meal (m) – (coarse flour) flat bread, porridge, etc.;
- grits (g) – (cracked and/or dehulled grain) brewing, intermediate processing stage;
- dehulled whole grain (w).

Symbols (letter in brackets) could be added as a suffix to the above quality to indicate particle size of milled product (i.e. Qc(m), meaning sifted meal and Qb(w), meaning dehulled whole grain).

As mentioned, an important factor affecting milling quality is fermentation. It can be carried out before milling, during intermediate stages (grits) and after milling. In the latter stage it must be considered a part of the food-processing technique. The capital letter F could be added as a prefix to show that the grain is fermented before milling (i.e. FQa(m), meaning fermented whole-meal coarse flour).

Furthermore, an important aspect in determining the shelf-life of flour is whether it has been milled dry, moist or wet. Sometimes it is necessary to add a small quantity of water to moisten the grain in order to soften the outer layer to enhance dehulling. The milled product must then be dried before storage. This is usually accomplished partly from the heat from the milling process and on mats where a thin layer is exposed to the sun. Wet milled flour must be consumed within a day. The capital letters D (dry), M (moist) and W (wet) can be added as a prefix to show these treatments, i.e. D(F)Qb(m), meaning dry-milled, fermented and partially sifted meal.

With some grain it is common to sieve the flour by hand before use in order to remove coarse fibres in whole-meal flour. This will increase digestibility and subsequently the quality of the milled product.

Any improvement of milled product can be indicated with a plus sign (i.e D(F)Qa(f)+, meaning dry-milled, fermented and hand-sieved whole-meal flour). Addition of soda might improve baking conditions but it reduces the destruction of phytic acid. In this case it might be feasible to add a minus sign to indicate risks for deficiency diseases.

The milling specifications will vary with crop, demand and use. It is the duty of the Technical Extension Agent (TEA) to discuss with home economics, nutrition and grain marketing experts, local millers and households before defining target groups and processing techniques. Table 2 is a summary of the common treatments discussed above in milling, and local conditions might add other alternatives. It is useful for the TEA to make his/her own local table of milling treatments to define local quality standards which should include both traditional and 'improved' processing.

Unfortunately there is a habit of many TEAs to start with a given 'improved' milling technique and overlook aspects such as demand and use of different milled products. Sometimes they can get along with it, but no doubt, the TEAs will achieve better results if they start with the different nutritional requirements for different target groups before looking into the marketing and the questions of technique. The choice of improved milling technique should be the result of a proper need assessment followed by a market survey of the demand for specific products.

As noted, vital nutrients are lost during the milling process. While the number of calories are almost the same for the three types of flour (Qa, Qb and Qc), the proportion of important nutrients are generally higher for whole meal (Qa) compared to Qb and particularly to Qc. This is especially true for calcium, iron, vitamin B and the fat content of the meal. Should adults rely exclusively on maize, they will have to consume 2–7 times more degermed meal than whole meal in order to satisfy their daily need of iron and vitamin B. On the other hand, both grinding and sifting is vital for improving the uptake of nutrients from cereals due to the negative effects of phytic acid, tannin, fibres, etc. No doubt, it must be the responsibility of the TEA in co-operation with home economics and nutrition staff to assist in defining the appropriate milling practices acceptable to the different target groups.

Table 2. Simplified flour treatment/quality chart		
MAIN QUALITY		**DESCRIPTION** Type of grain
Qa		**Whole-meal flour** (97-99% extraction rate)
Qb		**Dehulled and/or partially sifted flour** (80-90% extraction rate)
Qc		**Degermed, dehulled and sifted flour** (70-75% extraction rate)
GRADES		**DESCRIPTION**
PREFIX	SUFFIX	
	(f)	**flour** (cake flour, flat bread, etc.)
	(m)	**meal** (coarse flour, porridge)
	(g)	**grits** (brewing, intermediate processing stage)
	(w)	dehulled whole grain
(F)		**Fermented before milling** (soaking included)
(Mt)		**Malted** (sorghum, millet, etc.)
(P)		**Parboiled** (rice, wheat)
D		**Milled dry**
M		**Milled moist**
W		**Milled wet**
	+	activity to improve nutritional value of flour; i.e. hand sieving to remove fibres
	–	activity to reduce nutritional value; i.e. soda improves baking but reduces destruction of phytic acid
D(F)Qa(m)+ means dry-milled, fermented, whole-meal coarse flour where the coarse fibres are removed through hand-sieving.		

Through the milling process the pulverized kernel is exposed to oxidation and particularly the oil in the germ becomes quickly rancid. Moisture content and temperature are other factors speeding up degradation of flour. Degerming in combination with strict control of moisture content (below 12 per cent) are standard practices to increase the shelf-life of marketed flour. Fermentation of grain before milling is a traditional method to slow down the oxidation process and to improve the shelf-life of whole-meal flour. An important question for the TEA to look into is the benefits of improved shelf-life and for whom.

Recommendations on Flour Quality

The availability and abundance of food are vital factors for recommendations of appropriate flour qualities for the local diets. The better-off people are less susceptible to malnutrition while the poor rural and urban households are more exposed to nutritional deficiencies. Therefore it must be considered the duty of the TEAs to cater mainly for the less-fortunate groups.

Strong evidence speaks towards recommending fermented, dry-milled local grain where the coarse fibre has been removed as the most appropriate cereal product for urban and rural poor in Africa. The same nutritional value can be achieved through wet milling but the flour cannot be stored. The recommended milling technique should also enable the women to prepare specific nutritious food for children and pregnant women, such as malt porridge and germ gruel.

Large variations in food preferences exist and it is futile to adhere to rigid recommendations. It is the responsibility of the TEA in co-operation with home economics and nutrition staff to identify and promote the optimal acceptable solution for a given area.

ELEMENTS OF TECHNOLOGICAL CHOICE IN MILLING

3

CASE STUDY 1

Traditional Maize Processing in Eastern Province of Zambia, and Malawi

The common understanding of a 'traditional society' is a society where rules are set by the older generations. Changes occur, and if they are of a quite drastic nature, the society might be pushed into an unbalanced state of affairs which will require constant adjustments and management of crises. Today it is common that the younger and older adult generations differ in opinions on how to meet the imposed changes and it is usually the former who decide what to do. A recent study on milling in Zambia showed that the younger generation extension agents had a different perception of 'traditional' milling practices compared to the older generation of women in the rural areas. If the intention is to improve a technique, one has to start with a defined baseline technique and build from that level. In the Zambian/Malawian case it was obvious that one had to go back to the period before independence to find 'the real traditional' baseline milling technique.

This case study covers the eastern part of Zambia and Malawi and it goes back to the time before independence (1950–60). It will give a good understanding of the traditional baseline milling technique before the influence of mechanized milling and marketed sifted mealie meal (maize flour). This information is also highly relevant for Africa south of the Sahara due to the similarities of milling problems.

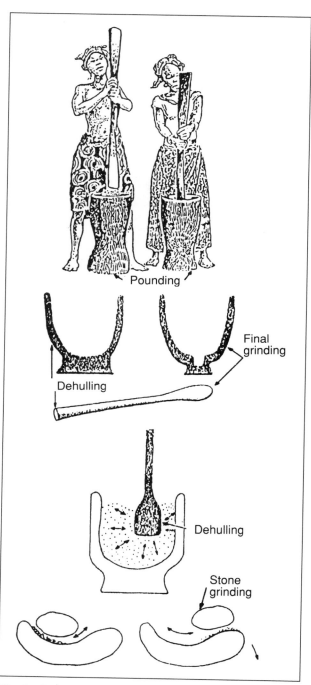

Fig. 3 *Traditional forms of hand-grinding*

Table 3. Manual processing of maize with fermentation in eastern province of Zambia and Malawi (traditional technology, pre-independence conditions)

PROCESS CHARACTERISTICS	PRODUCTS
STAGE I: DEHULLING Known as the 'first pounding' *(kukonola)*; the objective is to perform a clean separation of pericarp (bran) from the rest of the grain. (The rest of grain consists of the endosperm; the germ – which is rich in oil and minerals; and the aleurone layer – which is rich in protein, oil and minerals.) The pounding is performed with a pestle and mortar. Initially a pounding is created with a small amount of grain and water, and grains with poor 'poundability' may need sand added. When enough of the medium has built-up, pounding speeds up. Pounded mixture is winnowed in a flat sifting basket with the bran being shaken onto a mat. Fully separated whole grains are picked out, while grains with bran still attached are given a further pounding. About 12–15kg of shelled maize is pounded in one session to produce enough for 4–5 days for the average household. The task takes about 6 women-hours of physically demanding work, which may be shared. Some products are wasted, the percentage depending on wind conditions and the extent to which food is scarce. In contemporary conditions wastage is probably low, in the order of 3% (but can be much higher).	1. 'Crude bran' with high fibre content (i.e. pericarp with varying proportions of germ and endosperm). 2. *Mphale* is the main product; it comprises dehulled whole grain plus broken grains, germ and small pieces of pericarp. *Mphale extraction rates:* local varieties 78% ; hybrids 61–65% 3. After the first pounding some well-separated whole grains may be picked out and then boiled – often with beans producing a dish called *chingowa.*

STAGE II: TREATMENT OF BRAN
 	
The bran resulting from the dehulling may be entirely retained for consumption if food is in short supply, but usually a fraction of large flakes of tough and fibrous bran *(mankupeta)* is separated, using winnowing baskets and/or sieves, and fed to chickens.
 	
The remaining product consists of fine bran flakes together with quantities of germ and broken pieces of endosperm and is known as *madea or gaga.* The nutritional value of *madea* is related to the intensity of pounding and the degree of separation.The *madea* is sometimes subject to further separation to obtain a germ rich fraction *(mdzoole),* which is boiled to produce a highly nutritious dish (known as *mdzoole* or *phala la gaga).*
 	
Irrespective of whether *mdzoole* has been extracted, *madea* is kept moist for about 24 hours, during which time it ferments slightly. Fermentation improves taste, reduces phytic acid level and produces a stiffer porridge which is said to 'stay longer in stomach', allowing more complete digestion, giving more energy and postponing hunger feelings.
 	
The fermented *madea* is then dried, further pounded and sifted to produce a fine meal. This meal is then mixed with pure flour *(ufa)* in proportions which vary according to food availability. During the pounding/sifting of the fermented *madea* a further fraction is produced of small bits of endosperm, germ, and pericarp called *mitama ya madea.*
 	
Madea can also be used in the preparation of beer (although beer based solely on *madea* will be weak unless the carbohydrate content is high because of the poor pounding characteristics of the maize (hybrid maize).
	1. *Madea* is a fine bran meal (which is mixed with pure flour – ufa – in cooking a stiff porridge). The social status of porridge with *madea* is lower than that made with pure flour, but it is acknowledged to be more nutritious. Eaten frequently by women, especially during field labour peaks and when pregnant. Also consumed by men when food is short. Proportion of *madea* in the preparation served to men appear to depend on:
	
	    i) the general availability of maize;
	    ii) the household's income level.
	
	2. *Mdzole* when boiled this produces a 'germ meal gruel' high in fat and minerals.
	
	3. *Mitama* can be cooked as snack.
	
	4. *Beer madea* is an inferior brewing stock. Preference is to use a product higher in carbohydrates.

Table continued/

Table 3 continued

STAGE III: SOAKING	1. **Fermented whole grain** which compared to non-fermented grain has the following advantages:
The *mphale* is soaked in a large pot of water, the water may have been pre-heated (but this is not universal or essential and probably depends on fuel availability/prices). After 2 or 3 days grain is removed, and sufficiently pounded for the day's needs. The maximum soaking period is 5–7 days when fermentation is started in cold water. The water is changed at about 2-day intervals and the pot is washed, to prevent contamination of the grain by a sour taste.	* easier to pound (labour saving); * faster cooking (approx. halving cooking time); * superior digestability as perceived by customers; * preferred taste; * hygienic/better storage properties.
The process occurring during soaking is lactic fermentation which breaks down and acidifies the grain. Acidification has been demonstrated to kill harmful bacteria including those which survive cooking at boiling temperature; this is useful to human health and enhances storage properties of the flour subsequently produced.	
STAGE IV: FINAL MILLING	1. *Ufa* a product high in carbohydrate with low level of fat and minerals, and relatively low levels of protein. This is cooked to a stiff white porridge often made from maize flour alone (*nsima*). Local variants include mixtures of flour of sorghum, cassava and millet. (Maize *ufa* is also mixed with fermented bran flour – *madea* – described above).
Soaked (fermented) grain is washed until clean (normally three times). The grains are laid out to dry and then pounded in a mortar and sifted. Fragments which are resistant to pounding (*misere*) are retained.	
The pounding of 12kg of soaked grain takes about 7 hours spread over about 4 days (varying according to which fermentation has softened the grain). The final product is a fine white flour known as *ufa*.	2. *Misere* are boiled as a snack for women and children.
Extraction rates of flour (*ufa*) from shelled maize: local flint varieties 62%; hybrids (dents) 48-55% (mean 51%).	
Total time spent on Stages I–IV: 3-4 hrs/day per household	
Sources: Ellis et al, 1959	

Background

The early local maize in Zambia/Malawi was of a flinty type and was brought in by emigrants from the western part of Africa. Gradually through the mixing of dent maize from the southern part of Africa more semi-dent local varieties were developed. Most of the new improved high-yielding maize varieties today are of the dent types. The traditional maize-processing practice was developed around the flint maize. The new hybrid varieties were met with resistance by the subsistence households as they produced less flour through the local processing practices. Local semi-dent maize varieties are still common especially for subsistence use while hybrid maize is considered a 'cash crop'.

Traditional maize processing

Moistening and fermentation are important components in enabling dehulling, grinding, separation and conditioning of the main parts of the maize grain in the food processing chain. Figure 3 illustrates the use of the pestle and mortar for pounding and grinding with a saddlestone. Women and girls have the sole responsibility for the food processing. Table 3 describes the traditional maize processing in Eastern Province of Zambia and Malawi and the products achieved.

14

Comments

Table 3 illustrates a good traditional understanding of the nutrients in the maize grain, the skill to separate the main components and the knowledge of how to condition and prepare nutritious dishes for different members of the rural household. Through the fermentation process, the negative effect of phytic acid is reduced and thereby allowing for improved uptake of minerals, vitamin B and protein.

The experience from traditional maize processing casts doubt on the waste and nutritional irrationality arguments in rural food processing. It is certainly the case that the most visible end product is the very white flour, ufa, which has social prestige and is conventionally considered as the only suitable staple diet for men, or for entertaining visitors. When ufa is produced by village methods, about half the total crude protein and gross energy in the original grain is lost if the removed parts are not consumed. It is true that a fraction of tough bran is normally not used for human consumption, but when the dehulling ('first pounding') is performed on flints a clean separation is achieved and thus the loss of nutritious material from the discarded bran is probably low.

It is beyond question that in pounding of flint maize, the women are able to extract a number of other nutritionally rich products, including fine bran flour and germ meal (this is much easier with dents). Dishes prepared from these products are preferentially fed to children, and to women, especially during pregnancy and peak labour periods. Bran flour is also consumed by men during times of shortage. The practice of feeding a germ meal gruel to children assumes particular importance according to recent findings that malnutrition among younger children is related to the low energy-density of children's food like whole-meal flour.

The lactic fermentation which takes place during soaking, makes the grain softer and reduces the energy requirement in the subsequent milling, whether this is performed manually or mechanically.

The traditional maize milling is an extremely cumbersome processing practice and 3–4 hours of hard work per day throughout the year makes pounding an enormous task to perform. This is a good indication that, despite the hard work, the nutritional aspects of the traditional processing must have been given a very high rating in the evolution of the technique over the centuries. This reality ought to be an important factor when attempting to improve the milling technique.

Experience of sorghum and millet

The same traditional milling technique is applied on sorghum and millet and in addition to the mortar and pestle stone milling (saddlestone) is also practised for the final grinding. The dehulled by-product is used for beer production. Sorghum is considered more laborious to process traditionally compared to maize and it is common that the women will have to spend up to four hours/day to prepare the daily need of flour for the nsima (porridge).

CASE STUDY 2

Mechanized Milling of Maize in Eastern Province of Zambia, and Malawi

This case study is a continuation of Case Study 1 and forms the baseline information on mechanized small-scale milling. It describes the early introduction of hammermills based on pre-independence information from the eastern part of Zambia and Malawi. It shows the influence of the traditional processing on the early mechanized milling of maize. The information is highly relevant for the eastern and southern African regions in general terms.

Background

The traditional maize processing practice was developed around local flint maize. The new hybrid dent maize however, was met with resistance from the subsistence households as they produced less flour with the local processing practices. Already during the late colonial days, hammermills were considered important implements in popularizing hybrid maize as one could improve the extraction rate of locally produced flour from hybrids (the hammermill is described in Chapter 5). The combination of time saving on processing and improved extraction rate of flour were considered to be good arguments in the attempts to change from low-yielding local flint maize to more efficient hybrids. The persistent resistance to hybrid maize because of the processing weaknesses indicated that there was a need to integrate the hammermill into the traditional processing technique.

Mechanized maize milling with hammermills

Table 4 describes the experience of the early introduction of mechanized milling with hammermills. The data is unique as it integrates traditional practices with mechanized methods. Another unusual practice was to use the hammermill both for dehulling and grinding. During dehulling, larger screens were used in the hammermill to produce more coarse meal than during the final grinding. This meant two visits to the mill if mechanized dehulling was practised in those days.

Comments

The preferred products extracted from the milled grain were white flour (ufa), fine bran meal and germ meal. This could of course be achieved with the traditional practice but also through mixing of traditional and mechanized methods. The least-desired product among the rural households was whole-meal flour milled from unfermented grain (Option IV) despite the substantial time saving in processing. Based on existing experience, it appears reasonable to dislike Option IV as it reduces the possibilities to make nutritional food for specific needs (children and pregnant women). No doubt the best combined effects were achieved in Option I and II as they included substantial time saving, low milling charge (Option I), improved storage of flour and created more nutritional food options.

It is interesting to study how the practices have changed today, 30–40 years later. Traditional flint maize is still popular for subsistence use as it:

- stores better on-farm compared to hybrids;
- increases the possibilities for poor households to prepare more nutritious food for specific needs;
- requires less external inputs for production and processing.

Despite all efforts to promote hybrid maize it is still considered by many as a cash crop to be sold, processed and consumed outside the farm household.

Traditional dehulling and fermentation before milling in a hammermill is still a common practice in areas where maize is a traditional crop. Fermentation of the whole grain is also popular prior to milling in order to improve taste and digestion, increase shelf-life and to reduce energy requirement and wear in mechanized milling. Despite its common use, hardly any data exist on the use of fermentation today in Zambia or Malawi.

Whole-meal maize flour from hammermills is almost exclusively sieved before use in order to remove coarse fibres. Screen No 2 (2mm holes) is normally used to prepare a meal which can be hand-sieved before preparing the nsima, the local porridge. It is becoming a popular method to combine dehulling and milling particularly with hybrid maize. The dehulling technique with hammermills described in Table 4 is hardly used today.

STAGES	DEHULLING	TREATMENT OF BRAN	TREATMENT OF ENDOSPERM	FINAL GRINDING	COMMENTS:
Table 4. Option for mechanizing maize milling in Zambia/Malawi (pre-independence conditions)					
BASELINE	Traditional pounding, moist, winnowing	Traditional separation, fermentation	Traditional fermentation	Traditional pounding, moist	This is the traditional process described in Table 1 which produces the preferred white flour (ufa) and additional products.
OPTION I	Traditional pounding moist, winnowing	Traditional separation, fermentation	Traditional fermentation	Hammermill dry, screen No 2	Products are identical to the traditional process. Female labour is saved at a rate of 0.5hr/kg of shelled maize, minus time transporting to and from mill. Mill charges incurred in range of 1.5–2.5% of product value.
OPTION II	Hammermill dry, coarse milling screen No 3-4 hand sieving/ winnowing	Modified, not processed, fed to chicken	Traditional fermentation	Hammermill dry, screen No 2	Tough flakes of the pericarp will be detached from the endosperm during the dehulling/coarse milling process which will then be separated through manual sifting/winnowing. The extraction rate of mphale is normally higher than with traditional methods as a cleaner separation is usually achieved. Consequently, the bran has a lower nutritive value than the madea bran produced under traditional method, and is not normally processed further. Flour extraction is normally higher and flour has a slightly higher nutritive content as it contains material which is separated using the traditional method. However, the bran is the only by-product. Female labour is saved at 0.9hr/kg equivalent of shelled maize (plus time saved from not performing further processing of bran), less transport time for two visits to the mill. Mill charges are incurred in the range of 3–5% of product value (bran may be an element in paying).
OPTION III	Hammermill dry, coarse milling screen No 3-4 hand sieving/ winnowing	Modified, not processed, fed to chicken	No fermentation	Hammermill dry, screen No 2	This is an entirely mechanical process, except for the sifting or winnowing of bran after dehulling and the subsequent washing of grain. Dehulling normally reduces the weight by about 15-20%. Products: medium coarse flour which has not been soaked, this is known as mgaiwa, which is a generic term for village-produced flour which has not been fermented. This flour has higher shares of protein, fat and minerals than traditional ufa. But mgaiwa has inferior properties of storage, cooking, hygiene and taste and is perceived as less digestible.The bran is inedible and is not processed further. Mill charge is in the range of 3.5–7% of product value (bran may be an element in paying). Female labour is saved by cutting out fermentation, although this does increase the mechanical effort required for the final milling.
OPTION IV	Not performed	Not performed	Not performed	Hammermill dry, screen No 2	The whole grain is milled without the bran being separated out, producing a very coarse flour with no by-products. Being unfermented, it is also known by the generic term mgaiwa. This flour contains all nutrients available in the maize seed. But because it contains the whole milled pericarp, it has a lower nutrient composition per unit of weight than medium-coarse mgaiwa. It has the disadvantage of medium coarse mgaiwa above, and taste and digestibility are perceived as strongly inferior due to the high bran content. Mill charges in range of 3.5–7% of product value. Female labour is saved by cutting out fermentation, although this does increase the mechanical effort required for milling.
Sources: Ellis, et al. 1959					

Hammermills are in high demand and there are about 2,500 mills in Zambia today of which about two-thirds are in operation. The ability to pay for milling charges is an important factor in the choice between traditional or mechanized maize milling techniques. The capability to produce a marketable surplus is a determining factor for affording the milling charges and especially poor and sparsely populated areas have a low density of hammermills. However, the demand for the new technique is a sign that there is a change in preference/acceptability of flour quality.

Maize grits (maize rice) is a popular product for breakfast porridge and/or for several types of beverages. As they cannot be produced in a hammermill, the women have to use the traditional pounding technique.

The saving in hours due to mechanized milling is heavily dependent on the distance to the mill. A number of studies indicate that the introduction of hammermills for rural milling reduces processing times by at least one-half.

In rural areas where maize is not a traditional food and in poor urban localities, the rate of fermenting maize before milling is low. This is mainly due to lack of traditions, efforts to simplify milling, and marketing policies and efforts to provide subsidized sifted white flour from large-scale commercial mills all of which affects mealie meal preferences.

A recent survey in Zambia by the author shows that the promoters of small-scale milling, to a large extent, are men who have little interest and/or understanding of the benefits of traditional food-processing practices. They look at improved milling as a technical and economical problem, of which the main aim is to reduce drudgery for women. The principal product should be whole-meal flour. No doubt, these are important aspects but when fermentation is considered an outmoded practice despite its apparent benefits in the National Hammermill Project, concern is justified.

The present development trend indicates a need to revise the industrial approach to small-scale milling towards a farming system approach where the broader aspects concerning the use of cereal products will play a more prevalent role. Today there is a noticeable gap between the promoters of mechanized milling and the rural and urban women concerning the quality of milled product wanted and required. This question demands urgent attention as it affects millions of poor people, and particularly children and pregnant women both in rural and urban areas. In fact it is a responsibility of the Technical Extension Agents in Small-scale Milling to develop a more active dialogue between home economics agents, nutritionists, millers, traders and users of milled products to ensure maximum benefits from cereal processing.

Experience of sorghum and millet

As mentioned in Case Study 1, the same traditional milling technique is applied on sorghum and millet and in addition to the mortar and pestle, stone milling (saddlestone) is also practised for the final grinding. The dehulled by-product is mainly used for beer production. The difference comes when mechanizing the milling process. The bran in maize is tougher compared to sorghum and millet brans which are more brittle. During the milling process in a hammermill the maize grains produce flakes of crude bran which can be separated through hand sieving while the bran of sorghum and millet gets pulverized and cannot easily be separated after milling. The tannin in the bran of red sorghum causes digestive complications.

The problem with mechanizing sorghum and millet processing is a major reason for the limited cultivation of these crops today. Maize is taking over even in traditional sorghum and millet areas with lower precipitation, possibly resulting in increased food insecurity. Improved mechanization of sorghum and millet is a key issue in getting these excellent drought-resistant crops back in demand in drier areas. It is particularly the dehulling which is the main constraint in mechanizing sorghum and millet milling.

White sorghum has low levels of tannin in the bran and can be consumed as whole flour. A serious drawback is that it is very much liked by birds. The quelaquela birds which are common in Zambia and Malawi can cause serious damage to a crop of white sorghum within hours. Red sorghum is not liked by birds to the same extent and therefore can be an important crop provided the dehulling can be improved.

Elements of Technological Choice in Milling

Milling enterprises

One can briefly distinguish between three main types of milling enterprises:

- homestead milling;
- customs milling;
- commercial milling.

Then there is a number of intermediate stages, i.e. dehulling at home and final grinding at a customs mill. Figure 4 exemplifies the mode of milling, product flow and cost structure.

Homestead milling

The common homestead milling technology in most developing countries is influenced by the traditional processing practices. It is normally the responsibility of women who have to spend several hours per day on processing the daily need of cereal products. The technique is simple, but the work is heavy and time consuming. There is hardly any need of cash and the process is flexible in extracting and producing a number of nutrient products for specific needs.

The usual first mechanization step promoted is the investment in hand-operated plate mills (Figure 19). However, the technique is relatively expensive per kg milled product, will reduce flexibility and has proved to yield only marginal benefits compared to the traditional technique of pounding. In most cases the women prefer to bring the grain to a customs mill instead of the laborious hand cranking of small plate mills. It is not unusual that women earn some money on pounding for other households when they are producing their own.

Customs milling

It is possible to achieve a substantial saving of time and drudgery through customs milling. Many rural households are willing to give up the long and tedious dehulling and grinding by hand at home in order to devote themselves to other useful activities. The common operation selected is the final grinding in engine-powered mills (diesel/electricity/ water/ wind). The technique is simple and moderately efficient but reduces many of the traditional options to

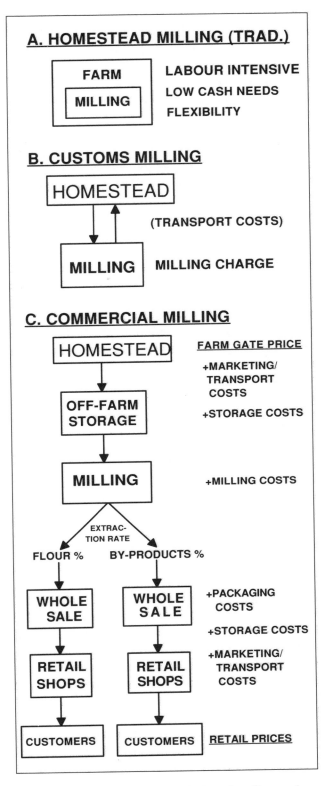

Fig. 4 *Mode of milling; simplified product flow and cost structure.*

produce selected nutritious cereal products. However, it is usual that the women perform dehulling, degerming, and fermentation before bringing the grain for final milling. In this way, one can combine the benefits of traditional and improved techniques.

A main disadvantage is the need for cash to cover the milling charges, which excludes many poor households. The time saving is heavily dependent on the distance to the mill, the mode of transport and the extra work needed to earn the money for the milling charges. The quantity milled beside available stock is usually determined by the quantity a woman can carry to the mill and this is seldom more than 2–3 weeks' needs. The requirement for specific considerations to improve flour shelf-life is of little importance for customs milling, as it is seldom stored for more than a few weeks.

There is a trend to increase the number of machines in customs mills to include also dehulling. This will expand the number of operations which can be performed. Improved dehulling technique is considered essential for increasing the production of more drought-resistant crops like sorghum and millet.

Many customs mills are also getting involved in small commercial milling, in order to increase the use of the expensive machines. The usual habit is to buy surplus grain and sell whole-meal flour to the local markets. This activity is common in customs mills in urban or close to urban areas. They sell the flour in small quantities and the question of improved shelf-life is seldom valid. If they can perform dehulling they will start to compete with the larger commercial mills on sifted flour.

Large-scale commercial mills

The complicated milling issues will arise with large-scale commercial milling. Figure 4 gives a simple product flow and where cost increases are involved. It is not the intention to describe large-scale milling but many of its aspects will affect small-scale milling. The demand for a marketed flour is a function of three major factors; consumers preference, price, and marketing. These factors are briefly analysed below.

Consumers' preference

It is argued that if consumers had a choice, they would prefer the high prestige fermented, degermed and sifted white flour in Africa rather than unfermented whole-meal flour. Surprisingly many consumers do not seem to be concerned – probably for lack of information – by the lower nutritional value of sifted flour.

In many urban areas where preference for sifted unfermented flour is established, the objective reasons for choosing this quality are because the flour has a more attractive appearance, better cooking characteristics and can be bought in small packages at food stores close by.

Among the subjective reasons given by urban consumers for their preference for unfermented sifted flour, advertising is by far the most important. It is evident that the market for this quality of flour has been heavily influenced by the technical nature of the larger rollermills. As fermenting complicates their operation they have, through advertising, presented unfermented flour as better. Government subsidies and their interest in large-scale milling has strengthened this development; for example in Zambia in 1987, the price of sifted mealie meal was lower than that of whole-grain maize!

It is obvious that a shift today to fermented whole-meal flour in many areas is a combined question of price, taste, attitude and information. The larger producer can afford to advertise but not the small customs mill. The latter would benefit from appropriate technical extension services and sponsored campaigns showing advantages and disadvantages of different types of flour. Dehulling service at customs mills can improve or reduce nutritional quality of flour depending on dehulling rate and marketing efforts.

Retail prices

Milling subsidies affect retail prices and they can distort the market. Differences in retail prices may also be a result of the following factors:

*   extra work with fermentation;
*   milling technique;
*   relation between extraction rate of flour and by-products. The price of the latter is generally much lower than that of whole-meal flour;
*   high packaging cost;
*   marketing system (transport, storage before and after milling, traders' mark-up, advertisements, credit, etc.).

The relatively higher retail prices of sifted flour do, generally, limit consumption to the middle-income and high-income groups in urban areas. In general an increase in price will force many to look for cheaper alternatives.

Geographical distribution

Rural areas consume exclusively the whole grain except for the coarse fibres. People do it traditionally through the separation of different fractions from which one can make different dishes. Whole-meal flour is gaining interest with mechanized customs milling. However, the inclusion of dehulling might increase the quantity but also the quality of more refined flour.

The situation is different in urban areas where both whole-meal and sifted flour are consumed by the urban population. In general the whole-meal flour is consumed by the low-income groups.

Sifted flour is appearing in areas short of grain or where cereals are not traditional. The main producers for deficit areas are the large-scale commercial mills using the local trading and/or emergency relief network.

Utilization efficiency

Milling technique will affect the availability of flour for human consumption. The extraction rate of sifted and degermed flour is lower compared to whole-meal flour, and the nutritious by-products from large mills are normally fed to livestock. This question is particularly important for countries in short supply of grain as they have to import more to offset the processing losses. No doubt, feeding animals is important but the conversion of bran and germs into meat results in a 90 per cent loss of nutritional value of these two by-products.

Given the deficiency of a number of essential nutrients in cereals, it is paradoxical that so many countries have allowed or even favoured the adoption of large-scale milling technologies particularly for urban areas which further reduce the amount of these nutrients per unit of output. The situation is further aggravated by the fact that fermentation is abandoned in this technology.

Composite milling

Bread and pasta made from wheat are becoming very popular in many developing countries where wheat is not a traditional crop and where it cannot be grown or grown only under irrigated conditions. Many countries have to import large quantities of wheat and it is common to talk about the 'wheat trap'. A number of developing countries are receiving wheat as aid on concessional terms which the governments then sell to the local markets, sometimes at a lower price than the local grain. This type of trade is an important source of income for many government agencies.

It is possible to substitute wheat flour with sorghum and cassava flour to a level of 20–40 per cent without an essential reduction in the quality of wheat products. The low wheat price does not favour the inclusion of wheat substitutes but another restricting factor is also the milling technique. If sorghum is used in composite flour with wheat it must be properly dehulled before co-milling.

Composite flour could be an interesting product for customs mills and small-scale commercial mills particularly in urban areas where wheat bread and pasta are popular products even among low-income groups.

21

PROCESSING REQUIREMENTS OF CEREALS, CASSAVA, LEGUMES AND CONDIMENTS

In this chapter the principal cereal grains, and cassava, legumes and condiments, are described, with a brief account of the milling procedures to be used. As mentioned earlier, the cereals are members of the grass family and Figure 5 shows the principal types.

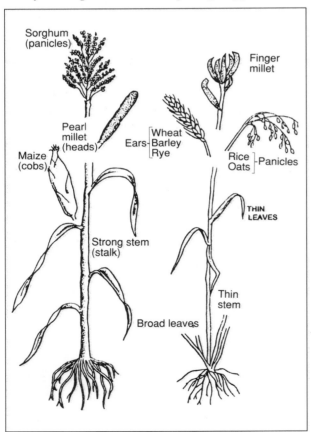

Fig. 5 *The cereal plant: principal types*

Wheat

Wheat is a crop of temperate and Mediterranean regions, grown both north and south of the equator. The principal world producers are in the developed countries but the crop is grown extensively elsewhere, including the high altitude tropics of Africa, such as Kenya and Ethiopia, north Africa, South America and south Asia.

The plant is a slender grass ranging from 60 to 250cm and with a compact head or `ear' which may be awned (hairy). The grain is threshed freely, without the husk. Wheat is principally used as human food (bread and chapattis) but in many countries surpluses may be used also as animal feed.

Grain structure and composition

The wheat grain is egg-shaped with a beard at the sharp end and has three principal parts:

* the germ, at the non-beard end of the grain;
* the starchy inner grain – the endosperm – is the main part of the grain;
* and the outer layers, collectively called 'bran', which cover the germ and endosperm and provides protection from pests and dehydration of grain.

Figure 6 shows the structure of a wheat grain.

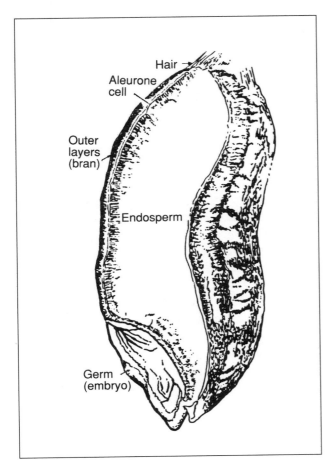

Fig. 6 *Diagrammatic longitudinal section of wheat grain through crease and germ.*

The germ is about 3 per cent of the grain by weight, and is rich in protein, sugars, fat and valuable vitamin B, as well as thiamine and vitamin E. Thiamine deficiency weakens the nerve system and eventually causes the sickness beriberi.

The endosperm is the part of the grain from which white flour is made. It is creamy white and makes up some 82 per cent of the grain mass. The bran is brown, tough in texture and rich in dietary fibre. The inner layer of the bran is the site of phytic acid, mentioned in Chapter 2, which restricts the absorption of minerals from the diet. For this reason, fortification of brown or high extraction flours is often recommended in commercially produced flour in developing countries (iron and vitamin B). The whole bran makes up about 15 per cent of the grain.

The varieties of wheat grown are often described as strong or weak. A strong wheat yields flour from which good bread can be made. This character of

strength in wheat comes from the protein, which must be of good quality and quantity. Weak wheats are ideal for making biscuits and cakes. Wheat that has been damaged by sprouting during a damp harvest will not make good bread, giving it a sticky crumb.

Wheat milling

In the milling of white flour from wheat, the objectives are to make as complete a separation of the endosperm from the bran as possible, and to reduce the endosperm to a fine flour small enough in particle size to pass through a 0.14mm sieve.

The wheat is first cleaned and then conditioned for milling by adjusting the moisture content, usually by adding 2 to 4 per cent clean water and allowing to stand. This changes the endosperm so that it breaks easily yet toughens the bran so that it peels off easily at the first stage of milling. As well as giving a good separation of bran, less energy is needed for milling if this conditioning process is used.

Hand pounding can be used for wheat as for other grains. The dampened grain is placed in a mortar and pounded with a heavy wooden pole. Output of flour is low in terms of time and effort.

The milling process can also be accomplished, as it was previously done in some countries, by means of pairs of circular, flat grinding stones, one stationary and one rotated, the two being in close contact. This system is known as a quern. The ground material issuing from the stone mill is a whole-meal flour. The stone mill can be adjusted in such a way that the endosperm can be released from the bran without being completely shattered. This enables the ground material to be sieved to separate some of the bran from the endosperm and so produce a whiter flour.

Plate mills and hammermills may also be used for small-scale production but these tend to break the grain into fine flour, making it difficult to separate the bran and endosperm.

The modern flour milling system is complex. It uses a succession of pairs of steel rolls (rollers) and a gradual separation of fine flour from bran. It has a relatively high throughput and is a high-cost operation which is not normally considered economically viable unless output exceeds 10 tonnes per day.

Customs mills grind the wheat to whole-meal flour, used for flat bread: these mills are usually stone mills or hammermills, running at 100-500kg/h.

Small-scale milling opportunities

For hand-operated systems there could be a lessening of manual labour if the quern were adopted instead of pounding. As hammermills and stone mills become more common for the production of fine whole-meal flours, improvements can be made by introducing sieving to remove bran. Stone mills give a very fine flour, particularly suited to chapatti making. Simple two-stage roller mills are available, operating at up to 500kg/h. These consist of a pair of break rolls and a pair of reduction rolls, with sieving between the rolls and after reduction. They are suitable for use in towns, where consumers or small shopkeepers can bring their wheat to be ground. Chapter 5 describes the range of mills available for wheat.

Maize

Maize originated in the western hemisphere and its cultivation follows the development of the extensive early civilizations of Central and South America where, to this day, it is the most important staple crop.

The maize plant is very distinctive, being a tall grass which, when mature, has a single strong central stalk varying in height between 60 and 500cm, with a number of tillers branching from its base. The grain, unlike other cereals, is contained in a cob, which consists of a central structure on which the seeds or grains are set in rows and is completely enclosed by a sheath of leaves. The length of the cob varies between 8 and 42cm and the diameter can be up to 8cm. Cultivars exhibit a wide range of colour in the seeds from white through light and dark shades of yellow, red and purple to nearly black but most staple grain is white to pale yellow. The two most commonly grown types are known as dent and flint maize. Flint maize, usually with a white endosperm, is a harder grain than dent and is more commonly used for direct human consumption whereas dent maize with a yellow endosperm is more commonly used in animal feeds. The advantage of flint maize is that it can be milled to produce maize grits which are preferred to flours in many food preparations. The relative hardness of

the kernel makes it more resistant to insect attack. Other varieties are grown for pop-corn and as sweet corn which is usually eaten as a vegetable.

Grain structure and composition

Maize is larger than other cereal grains such as wheat, though it resembles them in structure (see Figure 7). However, in maize the bran is hard and shiny, and is known as a hull. This hull is usually less than 2 per cent of the kernel weight. Immediately under this is the aleurone layer, which is a single layer of cells containing most of the seed's stored protein. The germ is larger than in other cereals and is usually between 10 and 14 per cent of the kernel weight and contains most of the grain's oil. The endosperm makes up the greater part of the grain, usually 80–85 per cent of the grain weight.

Milling considerations

At subsistence and village level the cobs are harvested by hand as they become mature. Some may be used immediately and are of special value in certain food preparations. The remainder will have their covering (sheaths) removed and then dried for long-term storage. Maize is stored both on the cob or as grain. Threshing of the grain from the central core of the cob may be achieved by hand-rubbing

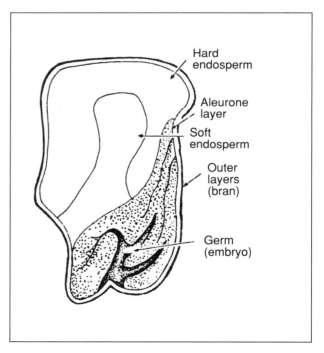

Fig. 7 *Structure of a maize kernel*

two cobs together is easier than picking out the grains with the fingers – by simple hand-held shellers or hand-operated rotary shellers. Mechanical shellers of the same basic principle as the last can also be used. Threshing machines are very efficient but the beating action – including the threshing mechanism in a combine harvester - can cause considerable breakage of grain.

Very few societies will consume grain without first removing the outer hull and various techniques are employed for this purpose. In most African villages, women spend several hours every day pounding, fermenting and drying grain for their families' needs. During pounding the grain is winnowed to remove the lighter flakes of husk and then continued until a meal or flour is produced. In Central America the maize is first boiled for a short time with wood ash and/or lime and then soaked overnight. This facilitates the removal of the hulls which may be achieved by subsequent hand washing and rubbing of the grain. The wet grain is then ground to a paste either by hand or by machine. In west Africa foods are prepared by wetting the grains and allowing them to ferment for a day or two before grinding in a plate mill. In eastern and southern Africa it is common to dry the grain after fermentation before final grinding in a hammermill.

Small-scale milling opportunities

For community use at village level, small machines are available for use as customs mills to process a few kilos of grain at a time or as a small commercial mill operating at 100 to 500kg/h. These may be simple hammermills with no extraction of hulls and germ, or hammermills combined with specific dehullers or small roller mills, having two pairs of rolls, which produce a fine meal, best described as granulated maize meal, free from hulls and of good keeping quality. Local preferences can be met in terms of fineness and separation of bran and germ.

At the home or village level, the separation of the germ and bran from the rest of the grain meal should not be encouraged without consideration being given to the nutritional consequences. However, sieving off coarse fibres is always a recommended practice. The difficulties associated with the development of rancidity during storage of processed meals does not arise in the family situation, because only sufficient grain is milled for the families' immediate needs. The bulk of the maize

stock is stored as whole grain, which will not deteriorate in the same way as processed meal.

If it is intended that a mill should serve the needs of a village or other small community, then a powered machine would normally be required. The choice would be, in increasing order of complexity and cost, between stone, hammer, plate and roller mills and dehullers. The choice of power between wind, water, electricity, and diesel will depend on availability, capital and running costs, and the ability of the user to run the equipment. These matters are discussed in detail later (see Chapter 6). The modern tendency is for the consumer or shopkeeper to pay for milling at a village customs/toll mill.

If a soaking or cooking process is used to loosen the husk, the grain can be milled wet to produce a paste: hammermills and stone mills are not suitable, so plate mills are used. The latter require the grain to be dried before milling. Although the primary need in the first case is for a mill capable of milling a wet grain, these machines can also be used to produce fine dry flours from grains or dried root crops by changing the metal plates to abrasive stone plates, thereby giving more versatility to the same machine.

In most places maize is ground dry and the three basic types of mill can be considered. With stone and plate mills the versatility of the machine and the fineness of the product is controlled by the distance set between the stones or plates. For these machines to be used efficiently, the mill operator needs to develop considerable skill to ensure good quality products without damage to the mill. Stone mills give a fine flour which is not required for making African-type porridges, though ideal for flatbreads. Hammermills are robust, require little attention, and once properly set-up are easiest to use and are probably best choice in the hands of unskilled operators. The particle size of the product is controlled by the mesh size (screen) fitted to the machine and largely controls the quality. In this way a meal or coarse flour can be produced which is exactly suited to porridges. The choice for much of Africa will be the hammermill.

Rice

Rice is the most important crop of the tropics and of the developing countries. Though considered as the principal food of Asia, it is important in Latin

America and increasingly so in Africa. Its evolution has given it tolerance to variations in water level and to competition from weeds, insects and disease. Such characteristics allow dependable yields under subsistence management. Rice culture can be divided into four types: irrigated lowland, which may be transplanted from nurseries or sown directly; rainfed; upland, which is rainfed and grown under the same conditions as other cereals; and flooded, where rice is grown in areas subjected to natural flooding to depths of up to five metres.

Traditionally, the rice plant of the tropics is tall (100–200cm) with long drooping leaves, and the grain hangs in a long panicle from which the grain – known as paddy at this stage – can be threshed by beating.

Grain structure and composition

Rice, oats and barley are threshed from the stem (with a husk covering the actual grain). Both in the field and in storage this husk protects the seed from pests and diseases. The edible portion, obtained by removing the husk, is known as rice, and shares common features with other cereals. Figure 8 shows the structure of a rice grain. Rice is usually eaten as a whole grain, not as flour or meal, so that quite different milling systems are necessary compared to other grains. There are exceptions, where rice flour is used as a substitute for wheat or maize in the preparation of local porridges.

After removal of the husk, which makes up 16–26 per cent by weight of the threshed paddy, by a first milling stage, brown rice is obtained. This consists of three important components: bran (5–8 per cent), germ (2–3 per cent) and endosperm (89–93 per cent by weight, dry). This is edible but does not store well; has a strong flavour, but is not easily digested. The bran and germ are therefore removed in a second milling stage to give white rice, which is preferred by the consumer, even though brown rice would be more nutritious.

Milling considerations

Processing of paddy to white rice is known as milling even though the product is not a flour or meal. Whether hand pounded, milled in one-pass huller mills or processed in modern multi-operational mills, the process consists of the two stages of dehusking and whitening.

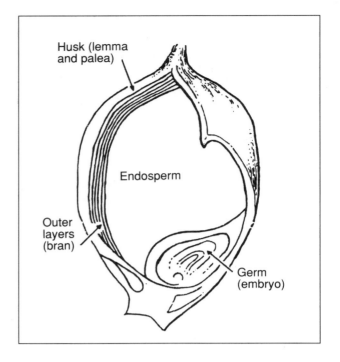

Fig. 8 *Rice*

In grain pounding, either hand or mechanized, the husk is split and removed by impact and any bran/germ removal is likely to be effected by abrasion between husk particles and the grain. In one-pass mechanical hulling, for which the brand name 'Engleberg' is used, husk is removed by shearing the paddy between stationary and rotating parts and further bran and germ removal is a combination of husk particle abrasion and inter-grain friction. The rate of grain breakage is likely to be excessive when pounding and steel hullers are employed and the milled by-product will be a mixture of husk and bran.

The multi-operational system includes discs, rubber rolls and centrifuges to remove husk; separating and grading devices, abrasive coated conesand cylinders, and improved friction machines to remove bran and germ with a minimum of grain breakage. At the large-scale mill, rice is cleaned before milling by a combination of aspirators, screens, specific gravity de-stoners, and size/shape graders to remove immature and diseased grains and foreign material. However, lower levels of process hygiene encountered in rural areas result in the majority of consumers washing their milled rice before use. This habit unfortunately reduces the nutritional value by washing out many of the vitamins.

Most milled rice (either raw or parboiled) is consumed as cooked grain. Variations occurring in methods of cooking include pre-soaking, boiling by total absorption or in excess water, steaming and addition of oil or fat to the rice and frying.

The by-products of rice milling are husk and bran including germ. Their value is related to their degree of intermixing, ranging from total use as a fuel source through animal feeds to extraction of cooking oil.

Small-scale milling opportunities

The need to maintain unbroken grain makes paddy processing particularly difficult. Simple hand-operated hullers are available but have relatively low throughput and are hard work, as are traditional pounding systems with mortar and pestle. The foot-operated *dheki* (see p. 36) is more efficient than pounding but is used almost exclusively in south Asia: it would be a useful addition to the technologies available in Africa, not only for rice but for other pounded grains such as pearl millet. Centrifugal dehuskers are less arduous in operation, but the dehusked grain still requires whitening.

For parboiled paddy, the Engleberg steel huller is suitable, and has the advantage that it is cheap to purchase and operate, and requires little skill. It is common in west Africa, and, as in south Asia, in both scale (half to one tonne an hour) and in socioeconomic terms, fits well into the village and small town environment as a customs or semi-commercial mill.

For raw rice, the modern rubber roll husker coupled with an abrasive whitener is a suitable system for greater than half a tonne an hour, though relatively expensive to purchase and maintain. The technology is available and already extensively used in Asia: it is appropriate only where, as is increasingly the case, a discriminating market demands high-quality rice and is prepared to pay a premium. The different steps of rice milling are illustrated in Figure 8a.

Sorghum

Sorghum and millets are drought-resistant cereals grown in the semi-arid areas where other grains cannot produce a guaranteed yield. As populations increase more of the drier marginal lands will be brought into use, with sorghum and the millets becoming increasingly important. In 10 African countries over half the daily calorie intake comes from them, and they are also important in India and China.

Sorghum is a strong-growing plant, with thick stems from 100 to 400cm high, bearing, in season, panicles of small seeds, which are the favoured food of weaver birds. Dark-coloured varieties have been evolved which are bird resistant. Unfortunately the red or brown pigment of such varieties is bitter and makes the grain less palatable and less nutritious until the outer layers containing the pigment have been removed.

Grain structure and composition

Sorghum grain varies greatly in size – from 2mm to 5mm diameter – and in pigment content. White and yellow varieties, frequently quite large, are highly desirable as food to birds compared to the red and brown varieties. A complete range exists between these extremes.

The kernel is typical of cereals (see Figure 9) consisting of outer covering, embryo and the storage tissue. Many varieties have an inner floury endosperm surrounded by a horny layer. This horny endosperm causes the grain to be very hard but brittle, which means that, compared to maize more energy is required for milling and it is difficult to produce a fine flour.

In addition to their bitter taste, the tannins in the outer layers of the grain also make it more difficult to digest the protein and are usually removed before the grain is eaten. This is done by hand pounding and fermentation or the use of abrasive machinery. Pounding in a mortar is carried out with the frequent addition of small proportions of water, thus loosening the outer layers, which are removed by one grain being forcibly rubbed against another under pressure of the pestle (pole).

Small-scale milling opportunities

Two principal products can be produced: a wholemeal flour or an extracted flour with some or all of the bran removed. The former is made by milling

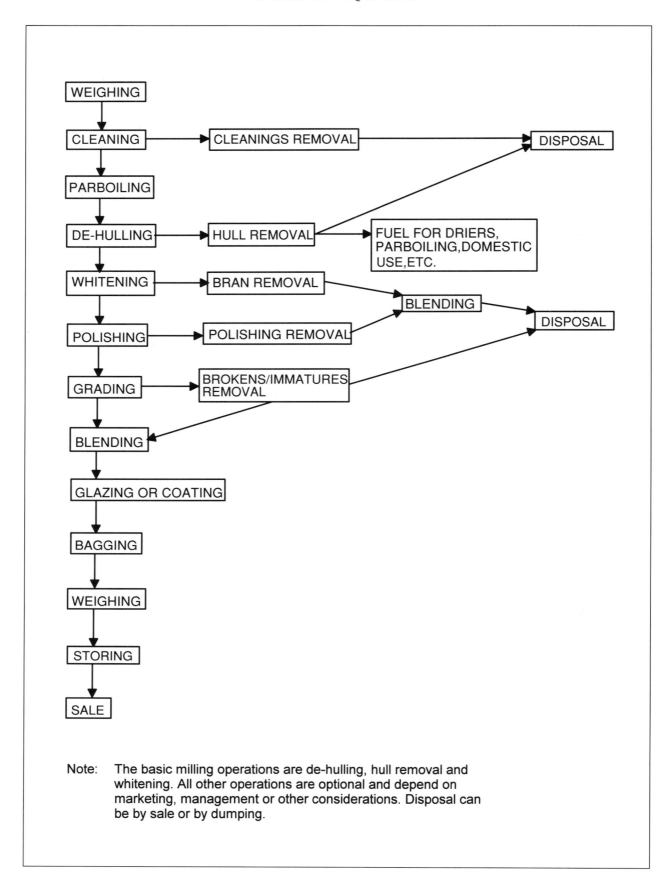

Fig. 8a *Rice milling flow chart*

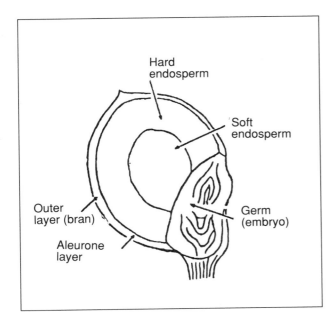

Fig. 9 *Microscopic section of a mature sorghum kernel*

cleaned whole grain to whole-meal flour and the latter is either produced by milling grain that has had the bran removed known as pearled grain or by sifting the whole-meal flour after milling. Pearled sorghum is made by removing the outer layers from the sorghum by abrasion, the scraping or wearing away of the outer layers of the grain by equipment specially designed for this purpose. These machines are a recent introduction and consist of rotating carborundum stones which rub off the outer layers of the grain (see next chapter). The pearled sorghum may be used as a rice substitute and as a more acceptable food than pigmented grain. Hammermilling will convert this pearled grain to flour or meal.

Traditionally, whole-meal flours are made by pounding or the use of grindstones, either hand or animal driven. Fine whole-meal flours are now produced in modern hammermills and stone mills driven by electric or diesel power. These mills are frequently operated as customs mills.

Some communities prefer the acid flavour of a fermented product. The sorghum may be wetted and allowed to undergo a fermentation, usually lactic, which not only modifies the flavour of the milled flour but also loosens the bran. Subsequent pounding in a mortar enables easy separation of the red outer layers, yielding a pale and pleasantly flavoured product. Mechanical milling can be used

successfully for this product. Germinated grains are used as sorghum malt, the starting point of traditional opaque beer (*chibuku,* etc.). After germination and drying, the roots and shoots are rubbed off and the grain milled. Again, this may be done in a mechanical mill.

Millet

There are many species of millet, but the most common in Africa is pearl millet, sometimes called bulrush millet, which is grown in the hot dry plains where other crops cannot provide a guaranteed food supply. Unlike the other millets, which have small grassy plants, pearl millet has long (up to 2.5m) thick (up to 3cm) stems, and long, broad leaves which are frequently bluish. Thus, its habit of growth resembles sorghum and maize. Millets are, of all cereals, the most resistant to arid conditions, and though yields may be comparatively low, only millets can grow on land of low fertility, intense heat and low rainfall.

Finger millet, *fonio, teff,* and other species are also grown. They are relatively easy to husk and grind. Finger millet is especially rich in minerals and is often added to other cereals before grinding. In addition, it may be germinated to produce a very sweet malt, used in weaning foods, in opaque beer and as a sweetener.

Grain structure and composition

The structure of pearl millet is similar to that of sorghum (see Figure 9). However, the outer layers are thicker, and the relative proportion of germ to endosperm is higher. The endosperm cells of millet contain greater quantities of protein especially in the floury endosperm area. As with sorghum, the outer bran layer contains a group of compounds known as tannins which impart bitterness in some varieties. Even within one variety there is frequently considerable variation in grain shape – from almost spherical to sharply pointed. Some variation in colour also occurs, though the colour is never as intense as that of sorghum.

Though it threshes free from husk, finger millet has a hard brown skin firmly attached to the grain. Other millets thresh easily, though some have a rice-like husk which must be rubbed off before milling.

29

Milling considerations

Before milling, the grain is cleaned by hand-winnowing and sieving. Because millets are very small, most dirt and straw can be easily sieved or winnowed out. On an industrial scale, mechanical grain cleaners would be suitable. As with sorghum, two principal flours are produced: whole-meal and extracted. The particle size and extraction rate are predetermined by the intended end use.

Traditionally, hand-pounding with the pestle and mortar is used to dehull and then grind the grain to meal. The saddlestone and the quern could also prove suitable, though not traditional in millet growing areas. Women carry out the milling domestically and, being the users of the flour, are in complete control of the process. The use of mechanized grindstones and hammermills is becoming more widespread, operating on a customs mill basis. As with hand-milling methods, the process may be adjusted to produce a range of products from coarse meal (for thin and thick porridges) to a very fine flour (for flat breads). For the latter, stone mills are more suitable than hammermills.

Barley and Oats

Barley and oats are grown in temperate zones north and south of the Equator. Oats can tolerate very wet and cold climates.

The principal uses of barley are for animal feeding and for malting and the brewing of beer. The usage of barley for human food (other than beer, which is not a subject for consideration in this review) is relatively small in the developed countries. Oats are used nowadays principally for animal feed in northern countries and the Andes of South America. There is still some use of rolled oats and oatmeal for making porridge.

Barley and oats thresh with the husk and are dehusked in machines similar to rice hullers. For brewing the grain is germinated in the husk. Barley malt is a major item of international trade for use in beer manufacture in countries which cannot grow barley. However, it has been found that sorghum malt is a suitable substitute.

Cassava

Cassava is a native root crop which is highly tolerant to drought and can produce acceptable yields on sub-fertile and acid soils. It is propagated from stem cuttings and has now become a very important food and industrial crop in many developing countries.

The roots of cassava do not require harvesting at any specific time of the year, but once mature and filled out, they can be left in the ground until required.

Harvested roots are highly perishable since they contain about 60 per cent water and have to be eaten within one or two days or processed to make a dry product that can be stored for a long period.

Many cassava varieties contain significant quantities of a toxic element (poison), cyanide. This toxin is naturally present in the roots and other parts of the plant, possibly as a protection against pests and other plant diseases found in the fields. However, the roots are processed before consumption to remove the toxicity.

Structure and composition

Cassava is a perennial crop which produces swollen roots (see Figure 9b). Many different varieties exist which produce roots with various shapes and sizes. The surface of the root is covered in a thin paper-like brown bark which is easily rubbed off with the fingers. Beneath this layer is the peel or skin. The peel is several millimetres thick and can vary in colour depending on the variety from deep pink to yellow to light cream. It can easily be removed in fresh cassava to leave the main bulk of the root.

This consists of swollen cells in which the starch accumulated by the plant as it grows is stored. At the centre, running lengthwise along the tapering root are fibres. The roots, after removal of the peel and fibres, consist mainly of water (60–70 per cent) and starch (15–25 per cent). The protein content is low relative to other root crops at 1–2 per cent. The fresh root can contain significant quantities of vitamin C (up to 40mg/100g).

The high-cyanide varieties of cassava, which must be processed before consumption, are bitter. Other varieties like 'sweet cassava' have low and accept-

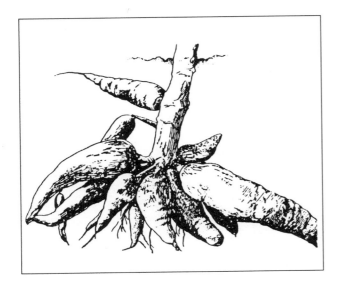

Fig. 9a *The underground structure of a cassava plant*

able levels of cyanide which make them suitable for direct consumption following normal cooking (i.e. boiling, steaming, roasting, etc.).

Processing considerations

There are many ways in which cassava roots are processed. The most common is by drying the roots to produce a storable product. Freshly harvested roots which show no signs of perishing are hand-peeled to remove the outer layers. If roots are left even for just a few hours and are allowed to dry out, peeling can become very difficult. The peeled roots can then be processed in a distinct way to produce dried products.

To dry cassava involves fermentation. The peeled roots are either left to soak under water until they have softened (3–5 days) or first grated and the mash placed into bags for 1–4 days. After fermentation, the soft roots or the mash are then dewatered by applying pressure to the material. This can be done simply by hand pressing, or by using heavy stones or a press to squeeze out the liquid. The pulp that remains can then be dried. This method results in a natural fermentation which produces an acid that binds cyanide.

The most common method of drying is in the sun. Pieces of the cassava pulp or softened root pieces are laid out in the sun, sometimes on specially made raised trays, sometimes on the roofs of hous-

es, or most commonly on a sheet on the ground or on rocks. The cassava is then allowed to dry naturally, which, depending on the season, can take 1–14 days. During the rainy season, drying is particularly difficult. The cassava has to be covered or brought in to the house each time it rains and the roots may never dry properly. This can reduce the length of time that it can be stored without going mouldy. An alternative to sun drying of cassava pulp is to dry the food using artificial means. This is often done by placing the pulp on a metal or clay tray, which is heated from beneath using wood or charcoal. The pulp then toasts or roasts, which drives off the water and causes the pulp to become granulated as the starch becomes gelatinized. This gives the product some special qualities, including the fact that the grains instantly absorb water and do not require further cooking. The product is known as *gari* in parts of west Africa.

The second method of drying does not involve an obvious fermentation stage. Peeled cassava roots are simply split lengthwise or cut into slices and sun dried. Because the roots have not been de-watered, they contain much more water and take longer to dry. During the process of drying mould may grow on the surface of the pieces. This assists in removing the cyanide toxin from the root pieces, but also causes the roots to become black and unsightly. After drying for 5–14 days, the root pieces are often stored in the eaves of the house, often above the fire where it is warm and smoky. This further assists in removing the remaining water in the roots, can inhibit infestation from beetles and allows time for the last residues of cyanide to be lost from the roots. Before the dry cassava is used for food, the pieces are scraped with a knife to remove the surface layer of black mould and smoke, to produce a whiter product. All of these dried products require milling or pounding into a flour before they can be used to prepare a food dish. Traditionally in Africa, the dried cassava, which is bought in the local market or taken from the household store, is pounded using a pestle and mortar. This is a labour-intensive task. More commonly nowadays, the village or market mills which are available for the milling of grains are also used to mill dried cassava. Small commercial mills buy roots, clean, peel, cut and dry them and mill the flour which is sold. In some countries grains and dried cassava are co-milled to produced an enriched flour.

The most common types of machine used for processing some of the common cassava products are small graters and mechanical presses. The graters consist of perforated metal cylinders cranked by hand or small motors. These increase the speed over manual techniques with which peeled roots can be grated prior to their further processing.

In the case of fermented products, removal of excess liquid from the mash assists in the speed and efficiency of subsequent processing and drying steps. Simple mechanical presses are used, ranging from balancing heavy stones on the bagged mash, constructing a counter lever stick press, or using car jacks to exert hydraulic pressure.

For the commercial production of cassava starch and tapioca balls, the peeled roots are wet-milled through a hammermill. The coarse fibres are then screened out and the liquor passed into tanks where the starch settles. This process takes two or three days. The water can then be run off, the starch dug out and dried in the sun. It is then hammermilled dry to give powdered starch.

Legumes

The term legumes covers beans, peas, lentils, groundnuts and many other species, some of which are trees and others small annual plants. They are grown mainly for cooking oil, flour and butter production or as vegetables.

Structure and composition

The seed is made up of an outer covering, the seed-coat, which is hard and when wetted becomes leathery; inside this is the germ and two seed pieces, known as cotyledons. All legumes have a scar (the hilum) where the seed breaks away from the pod.

Legumes are rich in protein, containing some 20 to 30 per cent dry weight and the rest as starch and fibre. They are usually cooked whole, but sometimes milled to flour, which is consumed either separately or used as extra source of protein in other dishes.

Processing considerations

Legumes are often milled by a simple grinding or hammermilling process, with no extraction of the seed-coat which is highly fibrous. There are two methods of removing the fibre. The first is to soak

the seed until the seed-coat is loosened and can be rubbed off. The seed is then dried and milled. The second method is to pass the seed through a break mill which splits the seed and free the cotyledons from the seed-coat.

Condiments

This term covers hundreds of spices, herbs and so forth, which are grown for their taste, smell, colour, or a specific nutritional or health attribute. The part used is often the seed but may be the root or stem or leaf. Condiments are used in small quantities, usually as powders in cooking but sometimes for drinks or chewing. As there are so many types, only a few principal ones are listed here with a brief statement on pre-treatment and milling. Salt is also a condiment. Domestic users are unlikely to take their materials to a mill. However, shopkeepers and other traders use either their own or a customs mill for grinding condiments. Some customs mills have a small stone or hammermill set aside for this purpose, as condiments are strong in taste and small, and sometimes corrode the machines. Otherwise, the hammermill used for grains will serve provided that it is cleaned well after milling condiments.

Roots: turmeric, licorice, ginger. These are carefully cleaned by washing, sliced and then dried. Shade-drying retains the flavour better.

Seeds: pepper, coriander, cumin, allspice, mustard, fenugreek.

Bark: cinnamon, quite dry when peeled from the tree. Is harvested when ripe and quite dry, but further careful drying may be needed before storage.

Dried whole fruits: vanilla, chilli, tomato.

Dried bulbs: garlic, onion. The fruit or the bulb is harvested ripe and dried, usually in the sun, but artificial drying may be used.

For all the above-mentioned, provided they are dry and brittle, milling characteristics are the same as for cereal grains, and hammermilling is suitable, using a screen of 1mm or less, so as to obtain the condiment as a fine powder. Stone mills may also be used, but are more difficult to clean and so remove the smell or colour of the condiment before milling the grain. The dried bulbs and chillies are sometimes preferred coarsely milled.

Dried leaves: fennel, wormwood, mint, thyme, balm, curry leaf, many localized types.

These are best dried in the shade to preserve flavour. Many have medicinal properties. Domestically they are pounded or hand-rubbed. Using the village customs mill or similar ones, they are coarsely milled, so a hammermill is suitable for large quantities, with either a very wide screen – say, 5mm – or no screen at all.

Salt

In some districts, sea or lake salt is used and normally stored as large pieces. These are easily milled in a hammermill to fine salt suitable for cooking and the table. Salt is highly corrosive – that is, rusts iron – and the mill must be thoroughly cleaned after milling salt.

Bone meal

Hammermills can be used for milling dried bones to bone meal provided they are cracked to smaller pieces by hammers before being fed into the mill. The bone flour is used as mineral substitute and local fertilizer.

Pre-treatments

Grain cleaning

Before milling, the grain (or other material for milling) must be cleaned. Grain from the thresher or threshing floor is often heavily contaminated with foreign material and imperfect grains. The presence of harmful levels of pesticides must also be considered.

On a small semi-domestic scale, hand-sorting or winnowing with a basket is very effective.

Foreign material

Foreign material acquired during growing, threshing, transport and storage, which is not part of the grain must be removed before milling. It can be of animal, vegetable or mineral origin.

Mineral origin

Contamination is a result of inappropriate handling, particularly during threshing, and is caused by inadequate technology, negligence and even intentional fraud. Examples are: stones, mud, sand, metal and glass, each of which can damage machinery and also greatly affect the quality of the product. Stones and mud are often picked up from the threshing floor when hand beating or oxen are used for threshing. Bread or porridge made from sand-contaminated meal has an unpleasant grittiness which wears down the teeth; stones in rice can break teeth. Small particles may also affect grain dryer efficiency. Odours and taints from, for example, urine, mineral oil or kerosene may originate from inappropriate shipment and storage conditions or from spillage in the farmyard.

Note: Definitions are given in the Glossary (Appendix 1)

Vegetable origin

Grain from the threshing floor will contain broken straw, husk (chaff), leaves, and other grains (weed seeds) which not only constitute an economic loss but also a potential source of infection of the stored grain by field pests or diseases. Particularly if the grain is not fully dried, there is a danger of moulds growing and spoiling the grain. Some moulds produce toxins, making it dangerous to eat the grain. Furthermore, dust from mouldy grain can also cause serious lung problems.

In cereals and legumes, individual grains and seeds do not mature simultaneously and at optimum time of harvest, a proportion may be immature. This is especially important for rice, where the presence of imperfect seeds can lower the value of the final product. Immature grains may have an 'uncharacteristic' colour and are often small compared with mature grains. In addition such grains are prone to breakage on milling and so will reduce the milling potential of the rice. Cracked grain can be caused by inappropriate drying or re-wetting of dried rice. These grains break on milling and thus reduce yield.

Animal origin

Foreign material of animal origin is evidence of poor food hygiene control, for example insects – dead or alive, insect fragments and droppings, rodent and bird excreta, and odours and tainting from these sources. This foreign matter is usually introduced during bad storage, and the contamination may lead to the whole batch of grain being made unfit for human consumption.

Pesticide residues

Pesticides are applied to crops to improve yield and quality and may be subdivided into three groups:

- seed dressings;
- agro-pesticides, in which treatment is applied to the plant;
- storage pesticides, in which treatment is applied to the grain and its processed fractions.

Husked grains have an advantage over naked grains in that residues of pesticides are likely to be removed or reduced substantially when the grain is dehusked during milling.

Peeling and drying roots

Cassava and other root crops are washed, peeled by hand and then washed again in clean water. If the product is to be a cassava chip, then the tubers are sliced by knife and spread on a clean concrete yard to dry in the sun. On an industrial scale, mechanical driers can be used, but, as two-thirds of the weight of the tuber is water which is to be dried out, sun drying is the more common method. Mechanical slicers can be used on a large scale. A full description of all stages of cassava processing, including *gari* production, is given in Chapter 4.

Legume splitting

The seed can be split by passing it through a bar mill or other break roller system. This is essentially the same as the break rolls of a flour mill, but the grooves on the rolls are larger. After splitting, the pieces of seed-coat are winnowed from the cotyledons.

Moisture and drying

Throughout the whole post-harvest system the moisture content of the grain is of critical importance and it is essential that grains and their processed products are kept dry to prevent or slow down respiration and the growth of insects and moulds. To remove excess moisture, sun-drying or artificial drying is used. Grain may be dried by spreading it in the sun on a clean surface, which may be concrete, tarmac, sacking, plastic and mats. In the case of rice, frequent turning is essential to prevent cracking of the rice kernel – shade drying is preferable to direct sun.

Artificial drying is essential in rainy harvests, when there is not enough covered over dry floor to spread the grain or sun to dry it. Mechanical drying is also preferred for large quantities of grain; driers may be continuous or batch. In either case, dry or heated air is passed through the grain to remove the moisture. The mechanical drying process is more easily controlled than sun-drying and, if land and labour costs are high, it may be cheaper.

Cleaning

The principles of grain cleaning are the same for both hand and machine cleaning. Sand and stones are heavy, but chaff and plant material are light compared to the grain that is required. The simplest form of cleaner is the winnowing basket, where the grain is thrown into the air and caught in the basket, the breeze blowing away fine materials and light plant materials. By careful use of the winnowing basket, stones and sand gather at one side and the grain at the other and eventually the grain can be collected and the stones discarded.

The same principle applies to large grain cleaners, where a powerful current of air made by a fan blows the grain and fine material away from the stones and then the grain settles first and the fine plant material is blown out of the machine. Such air systems should be augmented by a simple sieving system whereby grain and small material fall through a sieve leaving behind large pieces of mud and broken plant stalk on the sieve. A second fine sieve removes sand and dust from the grain and hand picking is needed to remove stones of a size similar to that of the grain. Machines usually incorporate both sieves and air classification systems.

Husk and Bran Removal

Manual husk/bran removal

PESTLE AND MORTAR

The pestle and mortar is the oldest and simplest method of dehusking and dehulling grain (see Figure 3).

For rice it is used primarily for dehusking and only partial whitening, as not all the bran can be removed by this system. Because high breakage occurs, it is more suited to parboiled rice with tougher grain. The rice is placed in the mortar and then the pestle is pounded (thrown) onto slightly moist grain. Sometimes two or three operators alternate with pestles to make the dehusking and whitening quicker. The impact not only breaks off the husk, but the stirring action under the pressure of the pestle rubs one grain against another, thus whitening the rice.

DHEKI

The *dheki* is a version of the pestle and mortar in which the pestle is at the heavy end of a pivoted beam. This beam is not counter-balanced and is raised by applying force downwards with the foot at the opposite end and then allowing the pestle to drop into the mortar under gravity. Being foot-operated, the *dhekis* are more efficient than the pestle and mortar (see Chapter 6 for a discussion of sources of power).

Capacity

Dhekis are made in the villages and homesteads to individual requirements and capacity will vary, though there can rarely be more than one or two kilograms of rice in the mortar at any one time.

Resource requirements

The pestle and mortar and the *dheki* are human powered, the dheki using the legs and the pestle the arms. Little skill is needed to make the pestle and mortar or the *dheki* and mortar, and there is virtually no maintenance.

Mechanized husk and bran removal

ABRASIVE DISC DEHULLER

This machine, like many others, uses carborundum stones or resinoid discs to abrade (wear away) the bran or hulls of cereals or grain legumes. Figure 10 shows a disc dehuller which is designed primarily to remove the bran from grains such as sorghum or millet, but it can be used for most grains.

Operation

In most cases the stones are rotated inside a metal housing. As the grain moves through the machine between the stones and the housing, the seed coat is

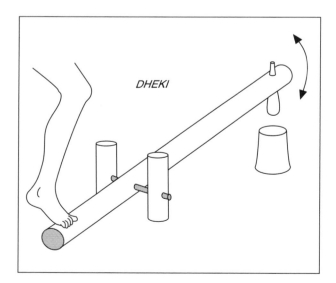

Fig.9b *A dheki*

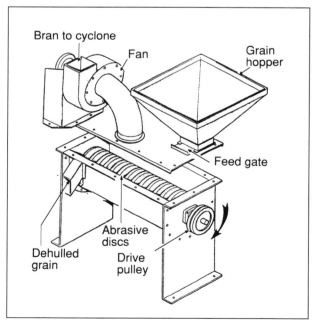

Fig.10 *Abrasive disc dehuller*

abraded. The stone discs may be set close together on the machine shaft to form a cylinder or may be separated as a series of discs.

The machne illustrated consists of a series of 13 fine grit carborundum stones set on a horizontal shaft at approximately 2.5cm intervals and separated by aluminium separators. The rotor so formed is mounted in a metal case lined with a heavy duty rubber coating. An aspiration system or cyclone allows the removal of the abraded material. This particular machine may be operated on either a continuous or batch method of use.

Capacity

Batch operation, at approximately a 5–6 minute operating time, 10–15kg (approximately 100–150kg/h). Continuous operation 250–500kg/h. Throughput is dependent on the hardness of grain that is being processed and the amount of dehulling required.

Resource requirements

The operator controls:

- speed of rotation;
- feed rate;
- disc gap setting.

Some skill is necessary, acquired by experience. Typical power requirement is approximately 8HP for 100kg/h.

This machine requires minimum maintenance. However, wear to the stones will occur and the operator should check this daily. Eventual replacement will be necessary, possibly every six months if the machine is in daily use for several hours a day. Improved performance is obtained by the occasional cleaning of the stones with a wire brush. Grain type and hardness will determine the lifespan of the stones.

STEEL HULLER

The steel huller, also known as the Engleberg huller, is used to dehusk grain which has the husk as a loose covering. The kernel must be fairly crush-resistant and the main use is for rice. The steel huller consists of a feed hopper, a cast metal ribbed rotor within a metal cylindrical milling chamber and a product collection/delivery device. An aspirator (see Glossary) may be included in some models.

The milling chamber contains an intruding metal blade to assist dehusking, and the bottom portion consists of a perforated screen to allow some venting of husk/bran particles. Figure 11 shows a section through a typical device.

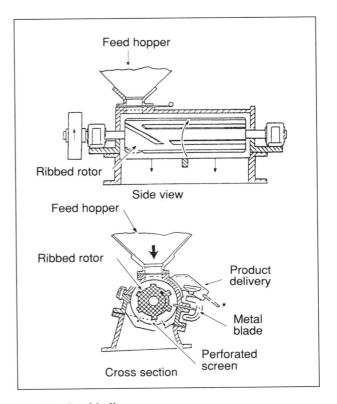

Fig. 11 *Steel huller*

Operation

The grain is screw-fed via a slide valve into the milling chamber. Dehusking is effected by forces generated between the lengthwise ribs of the rotor and the stationary blade, which shear-off the husk. Some whitening of the grain also occurs through rubbing the dehusked grain against the unhusked and therefore very abrasive rice. The necessary intergrain frictional force is generated by back-pressure from a slide valve at the exit of the milling chamber. The product mixture must then be separated, usually by aspiration or winnowing in small units. The by-product is a mixture of powdered husk and bran, which has reduced value as a consequence of its mixing.

To produce well-milled rice, an additional one or more passes through the machine may be necessary. To minimize grain breakage during multi-pass milling, the metal blade should be dominant in the first pass to effect dehusking; back-pressure, created by restricting the outflow of product, should be dominant in subsequent passes to effect bran removal. Machines have a slide valve to facilitate this restriction.

With maize, it is common to wet the outer layer before dehulling to improve efficiency and to reduce breakage of grain.

Capacity

Manually-powered devices are available at approximately 10–20kg/h. Motor-powered machines vary from 200kg/h upwards. These values are for one pass. Machines are usually driven by electric motor, but diesel engines are also used.

Resource requirements

The operator will be required to standardize grain size range, control the feed rate, and set the machine correctly for blade-to-rotor gap setting and back-pressure. These skills are readily learned.

In operation, dehusking efficiency is related to the condition of the rotor, blade, screen and valves, grain size and range, feed rate, back-pressure and rotor-to-blade gap setting.

Typical power requirement is approximately 1.3HP for each 22kg/h, so that a 200kg/h machine requires a 12HP motor.

The machine is extremely robust and replacement parts during normal operations are not required. Damage to screens, rotor, blade and valves can occur if the rice contains large stones, etc. The interval of replacement therefore depends to a large extent on grain cleanliness. Contamination with foreign matter affects wear rates and throughput.

Basic mechanical skills – i.e. ability to use spanners and screwdrivers – are needed to fit replacement parts.

CENTRIFUGAL DEHUSKER

The centrifugal dehusker has been produced in at least two designs, one operates in the horizontal plane, the other in the vertical. It consists of a feed hopper, rotary impeller, impacting target and product collection/delivery device. Figure 12 shows the operation in cross-section.

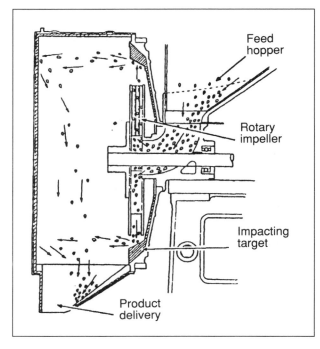

Fig. 12 *Centrifugal dehusker*

Operation

The centrifugal dehusker is used to dehusk grain in which the husk is a loose covering, not fused tightly to the kernel. Applications include rice, oats and sunflower.

The grain is fed to the centre of the rotating part (the impeller), and under the influence of centrifugally induced forces is thrown radially outwards. A ring-shaped target, usually coated in hard rubber or appropriate synthetic material, is located obliquely, that is at an angle, in the path of the thrown grain. On impact with the target, the husk bursts off from the rice grain. The product mixture must then be separated, usually by an aspirator.

In operation, dehusking efficiency is related to speed of impeller rotation and consequent impact force, grain size and range, and feed rate. The range of grain size (and weight) should be narrow for good performance. Grains at the larger end of the range will receive excessive acceleration and be likely to break on impact with the target; grains at the smaller end of the range will receive insufficient acceleration and retain their husk.

Capacity

A manually powered machine has a throughput of approximately 150kg/h. Motive powered machines vary from 250kg/h upwards.

Resource requirements

The semi-skilled operator is required to standardize grain size range, control the speed of rotation and the feed rate.

Typical power requirement for dehusking 500kg/h is 1.0HP; a husk separator may be required.

Target covers and impeller guides (that is grain throwers) will need occasional replacement: the interval of replacement depends to a large extent on grain cleanliness – contamination with foreign matter affects wear rates and throughput.

RUBBER ROLL DEHUSKER

The rubber roll dehusker is used to dehusk grain in which the husk is a loose covering, not fused tightly to the kernel. Applications include rice and oats.

The rubber roll dehusker consists of a feed hopper, two revolving rolls and product collection/delivery device; an aspirator is included in most models.

Figure 13 shows a cross-section of a typical device, and features the principle of operation.

Operation

The grain is fed to the nip (gap) between two rolls revolving towards each other on parallel axes. The rolls are coated in hard rubber or appropriate synthetic material, and usually are of similar dimensions. The principle of operation requires that they travel at differing surface speeds, so that shear

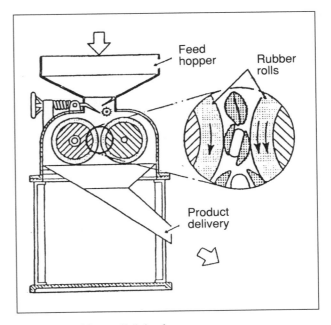

Fig. 13   *Rubber roll dehusker*

forces are generated and applied to grain passing between them.

The product mixture must then be separated, usually by an integral aspirator.

In operation, dehusking efficiency is related to the condition of the rolls, grain size and range, feed rate and roll gap setting. The range of grain size should be narrow for optimum performance. Grains at the larger end of the range will receive excessive shear and be likely to break on the rolls or to have their bran layers damaged; grains at the smaller end of the range will receive insufficient shear and retain their husk.

Capacity

Capacities vary from 200kg/h upwards, although most pairs of rolls in commercial use will dehusk about 500kg/h and require about 2.7HP.

Resource requirements

Operator skills, easily learned, include the ability to standardize grain size range, adjust the feed rate and roll gap setting and observe wear on the roll covers.

Replacement parts during normal operation are roll covers, usually changed by fitting a new pair of rolls and recovering the old rolls off the machine. The interval of replacement depends to a large extent on grain cleanliness. Contamination with foreign matter affects wear rates and throughput.

The rubber roller dehusker has performed below expectation in many villages in Africa. The main problem is the fast wear of rubber rolls and the problems to get replacement parts which tend to be very expensive.

UNDER-RUN DISC DEHUSKER

The under-run disc dehusker is used to dehusk grain in which the husk is a loose covering, although some abrasive whitening (scouring) of outer regions will also occur. Applications include rice and oats.

The under-run disc dehusker consists of a feed hopper, static upper disc and driven (rotating) horizontal lower disc, and product collection device. The discs are faced with an abrasive compound and set with a gap slightly less than the grain's maximum length. Figure 14 shows a section through a typical under-run disc dehusker, belt-driven from an engine or motor, to show the principles of operation.

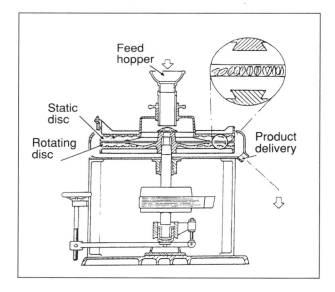

Fig. 14 *Under-run disc dehusker*

Operation

The grain is fed through the centre of the static upper disc onto the centre of the rotating lower disc. Under the influence of centrifugally induced

gravitational forces it moves outwards from the axis of rotation. The slight unevenness of the abrasive surface, the speed of rotation and the flow rate combine to tumble the grain as it passes between the discs. As the grain tumbles end-over-end in the restricted space, it experiences forces which shear off the husk. The product mixture is thrown out radially and must then be separated, usually by aspiration.

In operation, dehusking efficiency is related to speed of rotation, feed rate, grain size and range, and gap setting between discs. The range of grain size should be narrow for optimum performance. Grains at the larger end of the range will receive excessive shear and be likely to break; grains at the smaller end of the range will receive insufficient shear and retain their husk. Some damage to the grain's bran layers will occur in the dehusker, similar to partial whitening, and the grain should proceed to the whitening stage as soon as possible.

Capacity

Capacities vary from 500kg/h upwards. Typical power requirement for dehusking only is 4.4HP for the 500kg/h machine. A husk separator will also be required.

Resource requirements

There are no replacement parts necessary during normal operations. The surface of the abrasive compound (a magnesium oxy-chloride cement containing carborundum grit) needs occasional dressing/trueing and, longer term, recasting. The intervals depend to a large extent on grain cleanliness. Contamination with foreign matter affects wear rates and throughput.

The operator needs to standardize grain size range, adjust speed of rotation and feed rate, and check the disc gap setting. Skills can be learned on the job under knowledgeable supervision. Some precision is needed to dress and recoat the discs.

FRICTION WHITENER

The friction whitener is used to whiten grain: the primary application is for rice, where the kernel is fairly crush-resistant.

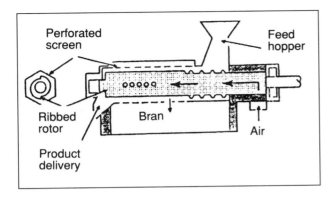

Fig. 15 *Friction whitener*

The machine consists of a feed hopper, a cast metal ribbed rotor within a perforated metal milling chamber, and product collection/delivery device. The milling chamber is aspirated to aid venting of bran particles and dissipation of frictional heating. Figure 15 shows a cross-section of a typical machine.

Operation

The grain is screw-fed via a slide valve into the milling chamber. Whitening of the brown rice is effected by intergrain frictional forces generated between the longitudinal ribs of the rotor and the static screen, enhanced by back-pressure from a weighted gate valve at the exit of the milling chamber.

In operation, whitening efficiency is related to the condition of the rotor, screen and valves, grain size and range, feed rate and back-pressure. Well milled rice may require an additional pass through the machine. To minimize grain breakage in multi-pass milling, the back-pressure should be reduced.

The by-product of whitening is commercial bran which also contains the germ plus any residual husk. Its nutritional value, and therefore financial value, is higher than the mixed product from the steel huller. Uses include animal feeds and the extraction of cooking oil.

Capacity

Capacities vary from 200kg/h upwards. Net capacity will be affected by multiple passes, but it is customary to have a row of two or three machines in series.

Typical power requirement is approximately 6.7HP for a 200kg/h machine.
Resource requirements

The operator is required to standardize grain size range, adjust feed rate with a slide valve on the feed hopper, and adjust the back-pressure with the slide valve on the product outflow. Experience is needed to balance these flows whilst maintaining product quality.

The machine is robust and replacement parts during normal operations are not required. Damage to screens, rotor and valves can occur if rice contains large stones, etc. The interval of replacement therefore depends to a large extent on grain cleanliness. Contamination with foreign matter affects wear rates and throughput.

General mechanical skills are needed to fit replacement parts.

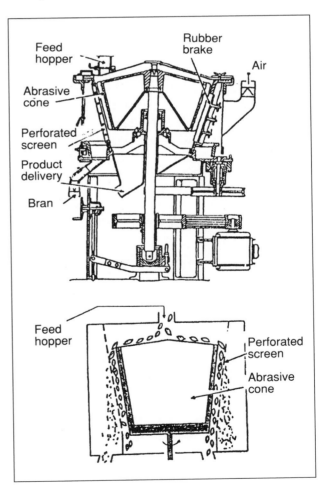

Fig. 16 *Cone whitener. Principle of operation*

CONE WHITENER

The cone whitener is used to whiten grain and the kernel must be fairly crush-resistant. The primary application is for rice.

The cone whitener consists of a feed hopper, an abrasive-coated conical rotor within a perforated metal milling chamber, and product collection/delivery device. The milling chamber may be aspirated to aid venting of bran particles and dissipation of frictional heating. Figure 16 shows a section through a typical machine with its own electric motor and the principle of operation.

Operation

The grain is gravity-fed via a slide valve into the milling chamber. Whitening of the brown rice is effected by abrasive contact with the rotor in combination with intergrain frictional forces, enhanced by grain acceleration/retardation induced by hard-rubber braking blocks intruding into the milling chamber.

In operation, whitening efficiency is related to speed of rotation, feed rate, grain size and range, and gap settings of cone-to-screen and cone-to-brakes. Well-milled rice may require an additional pass through the machine. However, the abrasive nature of the cone mill makes it possible to over-mill the grain and so reduce the yield of milled rice. Damage to screens, cone and brakes can occur if rice contains large stones, etc. The interval of replacement therefore depends to a large extent on grain cleanliness. Contamination with foreign matter affects wear rates and throughput.

The by-product of whitening is commercial bran which also contains the germ plus any residual husk. Its nutritional value, and therefore financial value, is higher than the mixed product from the steel huller, but is reduced by dilution with starchy endosperm from over-milling. Uses include animal feeds and extraction of cooking oil.

Capacity

Capacities vary, but are usually about 400kg/h, and for this a motor of at least 13.4HP is necessary, as the cone is very heavy and takes a high current at the start.

Resource requirements

The operator is required to standardize grain size range, control feed rate, and set the cone-to-screen and brake settings.

There are no replacement parts necessary during normal operations. The surface of the abrasive compound (a magnesium oxy-chloride cement containing carborundum grit) needs occasional dressing/trueing and, longer term, recasting. Rubber brake blocks also need occasional dressing/trueing. General mechanical skills are needed to fit replacement parts and to maintain operation and some precision is needed to dress and re-coat the cone. These dressing operations require skills learned by experience.

ABRASIVE WHITENER

The abrasive whitener is used to whiten grain; the kernel must be fairly crush-resistant. Primary application is for rice.

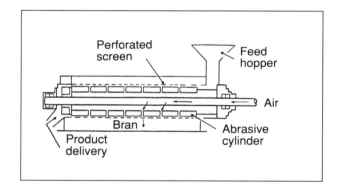

Fig. 17 *Abrasive whitener*

The abrasive whitener consists of a feed hopper, an abrasive-coated cylindrical rotor within a perforated metal milling chamber, and product collection/delivery device. The milling chamber is aspirated to aid venting of bran particles and dissipation of frictional heating. Figure 17 shows a section through a typical machine.

Operation

The grain is screw-fed via a slide valve into the milling chamber. Whitening of the brown rice is effected by abrasive contact with the rotor in combination with intergrain frictional forces, enhanced by back-pressure from a weighted-gate valve at the exit of the milling chamber.

In operation, whitening efficiency is related to the condition of the rotor, screen and valves, grain size and range, feed rate and back-pressure. Well-milled rice may require an additional pass through the machine. However, the abrasive nature of the mill makes it possible to over-mill the grain and so reduce the yield of milled rice.

The by-product of whitening is commercial bran which also contains the germ plus any residual husk. Its nutritional value, and therefore financial value, is higher than the mixed product from the steel huller, but is reduced by dilution with starchy endosperm from over-milling. Uses include animal feed and extraction of cooking oil.

Capacity

Capacities vary, but are usually about 400kg/h, requiring a motor of some 12HP.

Resource requirements

Operator skills are the same as for the cone whitener.

The machine is robust and replacement parts during normal operations are not required. Abrasive rotors for this type of machine are not normally dressed or recast by the user, replacement ex-supplier is the norm. Damage to screens, rotor and valves can occur if rice contains large stones, etc. The interval of replacement therefore depends to a large extent on grain cleanliness. Contamination with foreign matter affects wear rates and throughput.

Grinding

In the grinding process the grain is shattered and broken into ever smaller pieces until the desired fineness of flour or meal (coarse flour) is obtained. Pieces of husk/bran and germ can be sieved or aspirated (winnowed) out during the process if white flour or meal is desired.

Grinding mills may be driven by any type of power. Stone mills and plate mills can operate successfully at low rotational speeds, whereas hammermills must rotate quickly.

Manually operated grinders

The traditional methods of hand-grinding are very time-consuming and require a great deal of physical exertion. The saddlestone, for example, involves rubbing a small hand-held stone backwards and forwards over a larger stone to produce around 1.3kg/h of flour. Hand-operated rotary milling machines offer reduced physical exertion and increased output of about 15kg/h for coarse flour. Output drops sharply with fineness of flour.

Three factors must always be considered when building a milling machine to be operated by manpower (the term is used to mean human power - mostly, grain milling is done by women, who generate slightly less muscle-power than men):

- The strongest muscles should be used in order to produce maximum power;
- The speed of muscle motion should be kept in the order of 60–80cycles/minute without the need for an excessive amount of effort/work from the operator;
- The type of motion should be such that the operator is both comfortable and effective when working with the machine.

The thigh muscle is the largest and most powerful muscle in the human body. The usual pedalling/treadle speed of 60–80rpm uses the leg muscles at their maximum efficiency. Studies have shown that this speed yields approximately 75 watts, or one-tenth of a horse-power – twice the power of hand-operation. As with all machines, gears or pulleys may be used to alter the final speed of the mill stones to their best working condition. For milling, sawing and other similar tasks a flywheel is usually used to smoothen out the high and low power output of a person's natural pedalling rhythm.

Even a very fit person can generate at most about 50 watts with the arms and 100 watts with the legs, and this can only be maintained for limited periods. This limits the ability of manual power to deliver the quantities of milled grain needed to feed more than the immediate family and, even for this, the time taken can be undesirably long.

PESTLE AND MORTAR

This is the traditional method and is extremely labour-intensive. The separation of husk, bran and meal is relatively poor, and it can take several hours of pounding, winnowing, fermentation, drying and sieving to prepare enough meal for the family. Ease of hand-pounding has often been a main criterion in deciding which variety to grow. Traditional rhythmic songs may often be sung to ease the burden and monotony of labour.

The mortar is usually made of wood, but in some rural areas may be merely a hollow in the earth. The pestle is a straight, hardwood pole (see Figure 3).

The pole is thrown at the moistened grain in the mortar, using a slight spin. Efficiency is improved by using a large mortar and having two or more operatives, throwing down their pestles alternately. The spin of the pestle and the shape of the bottom of the mortar improves the removal of bran. In dry atmospheres, clean water, if available, is added during the process to assist bran removal and to reduce spill of ground material.

For pearl millet, only one-quarter of the effort is required for pounding off the hulls; the three-quarters for pounding to meal.

Experienced, fit and strong operators can pound 5kg/h.

Some studies have indicated that it is possible to increase pounding efficiency substantially by covering the point of the pestle and/or the bottom of the mortar with a harder material like metal. This can be a simple and cheap first step in improving the pounding technique, particularly if cast iron is used.

SADDLESTONE

The saddlestone has been in use for thousands of years and can produce fine flour for domestic use. It is very inefficient in terms of the human energy needed to grind one kilo of grain, as most of the energy is used to move the stone: the reciprocating motion is less efficient than rotary motion.

The grain is placed in the lower, saddle-shaped stone; the upper stone is then moved to and fro so as to crush and grind the grain between the two stones. Flour or meal is progressively winnowed from larger particles until the whole has been satisfactorily ground to meal.

The meal may be finer than with pounding, but the rate is only slightly greater (see Figure 3).

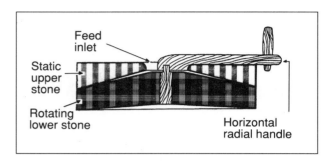

Fig. 18   *Hand-operated quern*

QUERN

This is the simplest rotary grinding mill, which is hand turned, and more efficient than the saddlestone. Figure 18 shows a section through a typical device. In terms of output and flour fineness the quern is more efficient than pounding. Its universal adoption has been prevented by lack of suitable stone in many places: the two stones should be hard yet easily split from the rock. Ideal rocks are very hard and with a rough feel, such as granite and sandstone.

Grain is fed through the hole in the upper stone while it is rotated. The grooves cut in the stones enable the removal of meal. These grooves are still of the same design used by the Romans 2,000 years ago. Their purpose is not only to enable removal of flour, but also to allow the grain to rotate, exposing a fresh surface to the abrasive stone. The flour falls from the edge of the stones into a trough and is periodically swept into a container.

Regrinding will be needed if very fine flour is required. Slight moistening of the grain eases separation of the bran, which can be winnowed or sieved out after a first coarse grind.

HAND-OPERATED ROTARY MILL

The rotary grinder is a simple plate mill, with a geared handle (see Figure 19). The plates rotate quite slowly as the handle is turned with one hand and the grain is fed into the hopper with the other.

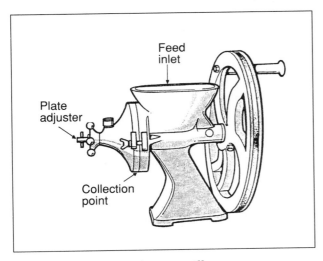

Fig. 19 *Hand-operated rotary mill*

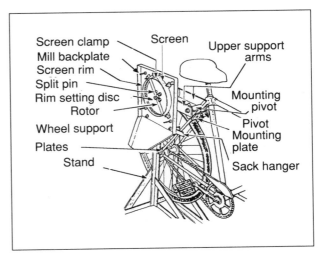

Fig. 20 *A pedal-operated grain mill*

The rate of feed and rotation are balanced by the user. Only coarse meals can be made. In Africa this type of mill, whether hand or mechanically powered, is used for the grinding of wet grain, as other mills are not suitable. For example, maize is soaked first and allowed to ferment for up to two days.

The fermentation softens the bran and also adds a sourness that is liked. The grain can be milled wet in the plate mill, but *not* in other mills, where the wet meal would block screens, stones, etc.

This type of mill can also be used for dehusking/dehulling of seeds such as sunflower and to produce grits.

PEDAL-OPERATED GRAIN MILL

Whereas efforts to harness pedal power to small machines such as lathes occurred long ago, it is only recently that efforts have been made to develop pedal-operated mills. An example of a pedal-operated grain mill is shown in Figure 20. The unit is made up of two distinct components: a grinding unit and a mounting frame – in this case, a bicycle.

The machine differs from other manual grinding machines in that it is a hammermill, in which the grain is broken down following impact with a rapidly rotating arm. In the more conventional hand-operated plate and stone mills, the grain is broken down between two closely spaced grinding surfaces.

The operator turns the pedals at a normal, brisk cycling speed. The cycle wheel in turn drives a roller shaft by frictional contact with the outer tread of the tyre at a speed of about 5,000rpm. A fixed rotor arm

is fitted to this shaft within the milling chamber and the grain is broken up after it is struck by the tip of the rotor. A screening mesh controls the fineness of the grinding process. By changing the size of the roller, different speeds can be obtained.

The milling process requires two operators: one to perform the cycling action whilst the grain is fed into the mill by a second operator, a handful at a time to provide a steady loading.

For a given effort, the output of the mill depends on the fineness of the product required. The mill works best on hard, brittle grains such as maize, millet and sorghum and on legumes.

The design of the milling attachment for a standard bicycle is kept simple to minimize manufacturing costs but where a bicycle is not available for this temporary modification or where a fixed mill is thought to be more appropriate then the cycle could be replaced by a simple wooden frame. The wooden frame would need to support the operator, a bicycle wheel and the pedal drive mechanism.

Despite its simplicity, this design has not been accepted in rural areas as a feasible option to mechanize milling. There are too many social/cultural barriers for women using pedal-operated grain mills.

Mechanized mills

There is a wide range of milling equipment and techniques available to match the needs of the processing scale and end-product requirements. The

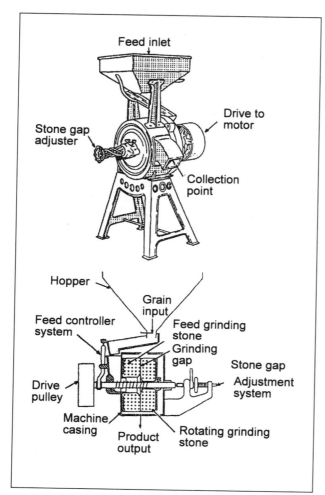

Fig. 21 *Diagrammatic representation of a mechanical stone mill with vertical grinding stones.*

smallest scale of milling is associated with grinding of cereals using a mortar and pestle, or stone querns turned by hand, whilst roller mills are used at the very largest scale of commercial milling.

In between these extremes, mechanically powered stone, hammer and plate mills are available to grind cereals and grain legumes to whole meal. Until the introduction of roller mills during the nineteenth century, all flours and meals were made by first grinding or breaking the cereal to a coarse meal. If a finer flour was required, this meal would be sieved to provide an extracted, finer flour. The difficulty in making a white flour in this way is that it is not possible to separate all the endosperm completely from the bran, or to separate the bran particles from the white flour if the bran has been reduced to the same size as the flour. Roller milling does permit a near-perfect separation by peeling the bran from the grain and separating it from the endosperm before the final crushing to a fine powder takes place.

STONE MILLS

The milling process is achieved by passing grain between the flat surfaces of two millstones, one of which rotates whilst the other remains stationary. A shearing action results which causes the grain to be reduced to a meal or flour. The advantage of stone milling over other forms of one-step milling is that a finer flour can be made than by other simple one-step processes.

In the older, traditional water- or wind-driven mills, large millstones, over a metre in diameter, were employed. Because of the weight, the stones would usually lie flat with the base or bed-stone fixed and the upper stone rotating. Naturally cut stone was commonly used in this type of mill. Today millstones are usually of artificial stone, manufactured from a composite material: often a mixture of emery and carborundum embedded in a matrix of magnesium oxychloride or, more simply, small pieces of natural stone and emery embedded in a cement matrix. The stones may be horizontal if they are one metre or more in diameter, or vertical if smaller. The rotating stone is driven by an electric motor or diesel engine. With vertical stones, a spring pushes the stationery stone towards the other. The gap setting can be adjusted to produce finer or coarser flour.

Although stone mills can still be arranged to be powered from a local stream or by wind, small commercial mills with electric or diesel motors are available and are more commonly used. In modern stone mills the millstones may be set either horizontally with a vertical rotary shaft, or vertically with a horizontal rotary shaft. The diameter of the millstones varies according to machine model type and size. Generally, because of the weight of the millstones and the relative difficulty in supporting them in an upright position, millstones used vertically are smaller in diameter (20 to 60cm) than those used horizontally (60 to 100cm). Figure 21 illustrates the basic design and principle of operation of a stone mill.

Operation

In a typical stone mill, a cone-shaped hopper holds the grain which enters the milling chamber through a feed control valve, usually of the simple slide-valve type. A shaking device often augments a gravity feed system and a screen prevents large impurities from entering the milling chamber and causing damage to the millstones. The grain from the hopper is fed, through the central hole in the rotating stones, into the gap between the stones. As the rotating stone moves against the stationary stone, the grain is ground or milled as it travels from the centre to the edge of the stone. Where the stones are set horizontally, the crushed grain is moved to the edge of the stones by centrifugal force, whereas gravity assists the movement of the crushed grain between the vertically set millstones.

The grinding surface of the millstones are dressed or intersected by a series of grooves, which assist the grinding action and help move the grain over the surface of the millstones. It is usual for millstones to be enclosed within a supporting and protecting metal band.

Capacity

The average output of stone mills varies between 25kg/h and 1,200kg/h, depending on raw material, fineness of grind, size of stones and motive power. The capacities of electric motor or diesel engine used to drive a stone mill will be from 1 to 27HP according to mill capacity and the diameter of the millstones: a typical mill grinding 500kg/h needs a 6.7HP motor. The motor should be geared to run from 400rpm for large horizontal stones to 800rpm for small vertical stones.

Resource requirements

Experienced and trained staff are required to operate this type of mill, to control feed rate and set and maintain the milling gap between the stones. Damage to both the stones and the milled products occur if the millstones are allowed to come together too closely or in contact with one another. The operator can check by feel if the flour is getting too hot – a sign that the stones are too close and the grind too fine.

As the millstones begin to wear and become smooth they will need to be dressed or reversed and some experience is required to do this and to reassemble the machine.

Maintenance and spares

Although this is a robust machine, damage to the stones will occur if the grain to be milled is not cleaned, nor stones and metal objects removed. Millstones, whether artificial or natural, wear

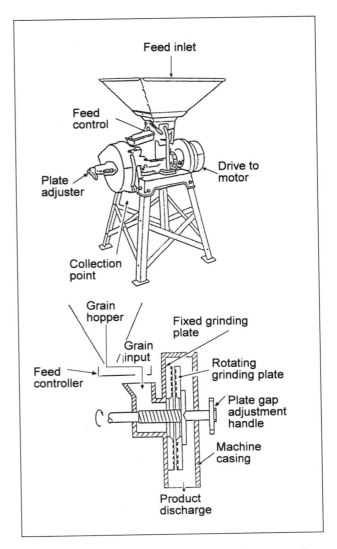

Fig. 22 *Diagrammatic representation of a plate mill*

smooth as they are used and will need to be first reversed and then, as wear continues, replaced.

PLATE MILLS

This type of mill is especially suitable for grinding wet material. In Central America plate mills are commonly used to grind wet maize to a paste (*masa*) for tortilla making. Figure 22 features the principle of operation.

Plate mills have been developed from the traditional stone mill with the millstones replaced by relatively small grinding plates made of cast iron, with grooves on both sides so that the plates can be reversed. The mill is designed and operated in much the same way as a stone mill: with one plate fixed while the other is turned by a belt driven from an electric or diesel motor.

Operation

The grinding plates are set vertically within an enclosed milling chamber. One plate is fixed while the other is turned at a speed around 600rpm. As already noted, there are small domestic hand-turned versions of this machine. The grain is either gravity- or screw-fed from a hopper into the gap between the plates. This gap may be adjusted to vary the fineness of the ground material and the grooves aid the shearing and grinding of the grain. Different plates, with a range of groove sizes, may be used to change the particle size of the ground material.

Unlike stone and roller mills, plate mills can be used to grind wet grains.

Capacity

Typical outputs for a machine with a 6.7HP electric motor:

- Wet Materials 200kg/h
- Dry Materials 350kg/h

Resource requirements

Experienced and trained staff are required to operate this type of mill, to control feed rate and set and maintain the milling gap between the plates. The plates may well be damaged if allowed to come into contact with one another whilst in motion.

Maintenance and spares

The plates will wear smooth as they are used and will need regrinding from time to time. To extend the period between regrindings the plates are generally reversible. As with other mills the life of the machine is prolonged if foreign matter of mineral origin is removed from the grain prior to milling.

HAMMERMILLS

The simplest form of one-step milling for dry cereals and grain legumes is the hammermill, used throughout Africa for the dry-milling of maize, sorghum, cassava, etc., and is often locally manufactured. Unlike the shearing action in a plate or stone mill, size reduction in a hammermill occurs principally by impact as the grain is hit by rotating hammers or beaters. Further impact occurs between grains and as they hit the casing of the machine and the metal

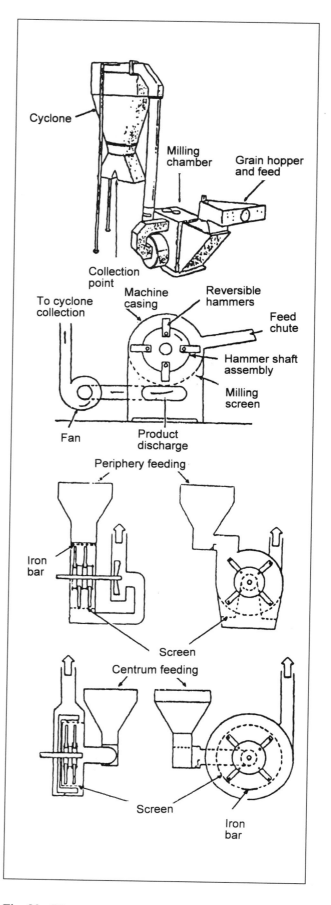

Fig. 23 *Diagrammatic representation of hammermill*

screen. The broken grain is retained within the milling chamber until its size is reduced sufficiently to allow it to pass through the screen perforations. A wide range of screen sizes is available for the production of different fineness flours and meals.

Figure 23 illustrates the basic design of a hammermill equipped with a fan to blow the ground material to a cyclone where it is collected in the lower part. One can distinguish between two different types of grain feeding into the milling chamber:

- periphery feeding;
- centrum feeding.

Operation

The design and capacity of hammermills vary between manufacturers. In general they comprise a metal body with a circular milling chamber through which passes a horizontal rotary shaft powered by an external energy source. Hammers are attached to this shaft and may be fixed or swinging. As the hammers need to rotate at high speed to cause the necessary impact and breakage of the grain, only electric, diesel motors or hydro-power plants can usually do this satisfactorily.

The hammers are made to rotate at speeds up to 3,600rpm. The fixed hammers are bolted to the shaft and best made from a hardened steel. The advantage of swinging hammers is that they will fall back if the chamber becomes overloaded and thus protect the motor.

Optimum periphery speed of the hammers is between 70–110m /sec. This means that small rotors must operate with higher speed per minute (rpm) compared to bigger rotors. An unbalanced rotor rotating at high speed will vibrate heavily causing premature damage of bearings and cracking of welds and metal parts.

A screen, mounted on a fixed circular support, surrounds the hammers. The grain must be sufficiently reduced in size to pass through the screen before it is discharged from the milling chamber. Most hammermills are supplied with a range of screens and unless a very fine flour is required, 1.5–2.5mm screen holes are suitable for most human foods (porridge) whereas 3mm and above produce grits and animal feeds. A hopper fixed above the milling

chamber holds the grain which is usually gravity fed into the mill.

A hammermill with peripheral feeding produces a meal/flour with finer particles (less than 0.2mm) compared to a mill with centrum feeding. The former type has a heavier rubbing action compared to the centrum feeding whereas the grains are thrown against the screen and break to pieaces. This means that the latter will produce a more even sized meal/flour with less fine particles below 0.2mm which is desired for tasty porridge and livestock feed. A mill with centrum feeding will also produce less dust and demand slightly less energy for milling a given quantity of grain compared to a mill with peripheral feeding.

The hammermill works best with grain of a water content of 12–14 per cent. The potential capacity increases with extended screen surface as there will be more holes for the flour to pass through. Increased distance between the hammers and the screen reduces capacity. That is why capacity goes down with worn hammers. Mills fitted with thinner hammers produce more even particle sizes of the meal which is an advantage for porridge.

When the hammers rotate in the hammermill they create a fan action which blows the flour out of the milling chamber. Figure 23 shows a hammermill equipped with an extra fan which will suck the flour from the milling chamber and a cyclone separator for bagging the milled product. The fan will improve grinding capacity by 25 per cent and increase transport flow of ground material to the cyclone. The purpose of the cyclone is to separate the flour from the air stream and to fill the attached bags.

The simplest hammermills have no cyclone or fan. The ground material is discharged by gravity through the screen directly into the bag. This is a simple but dusty system. A great improvement is the cyclone separator which reduces dust substantially and makes bagging easier. With a short distance between the mill and cyclone, the fan action of the hammers is normally strong enough to blow the flour into the cyclone if the capacity requirement is modest. When higher capacity is required the fan action from the hammers will be inadequate and therefore, an extra fan must be fitted to the main shaft. To a certain extent a larger engine can compensate for the extra fan action but this is a rather wasteful method to increase capacity.

In many developing countries it is a common practice to use simple designs of hammermills and compensate for output with larger engines. Under most village conditions a one-cylinder diesel engine of 7.5hp should be enough to cope with the quantity milled per hour. Usually it is more profitable to spend money on a good mill design instead of buying larger engines

A major question to look into is the quality of workmanship in the manufacturing of the mill which is exposed to vibration both from the mill itself and the engine. Pieces out of alignment, poor joints and weldings are strong indications that a mill has serious weaknesses and will not last as long as expected.

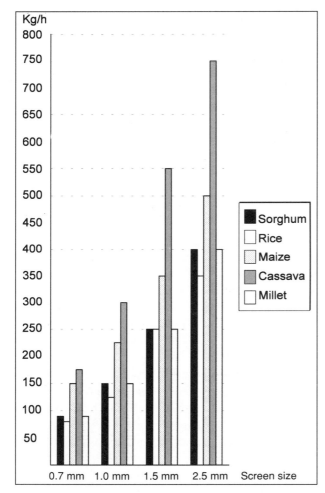

Fig. 23b *Relation between output of hammermill and screen size.*

The hardness of the hammers is a subject of regular discussion. The harder they are, the longer they will last but they will be more expensive. The mill will still produce meal with rather worn-out hammers but to what cost? The energy requirement can be doubled per kg milled product with worn hammers. Studies indicate that the optimum distance between hammers and screen should be around 8mm.

Capacity

The output of ground material varies according to the capacity of the motor, the size of the perforations in the screen and the variety and moisture content of the grain.

The following may be taken as a guide to outputs with different sizes of screens from a typical hammermill with an 8HP motor.

The data in Figure 23b is supplied by A/S Maskinfabrikken SKIOLD. It clearly shows the relation between screen size and output. A reduction of the common 2mm screen to a 1mm screen will reduce capacity by half and double the the power requirement and the operational costs. In general it is recommended to use a 2mm screen for a meal used for porridge. For baking flour it might be necessary to use a smaller screen. The selection of the right screen is an essential component for the milling economy.

Resource requirements

Experienced and trained staff with some mechanical ability are required to operate this type of mill, in particular, to check the raw grain for metal, stones and sand which can damage the screens, to control feed rate, check the screen and change for different products required by the customer. Most hammer- mills are run on a customs mill basis, so some financial training is necessary for the manager or principal operator.

Maintenance and spares

All types of mechanical grinders require regular maintenance if they are to perform the grinding operation efficiently. All moving parts should be lubricated regularly and screens inspected for damage. The hammers should be inspected for wear and sharpened where possible. Swinging hammers can normally be reversed and where necessary resharpened. They are easily replaced: the operator must be able to use ordinary mechanic's hand tools (spanners, etc.). As with other mills the life of the machine is prolonged if foreign matter of mineral origin is removed from the grain prior to milling.

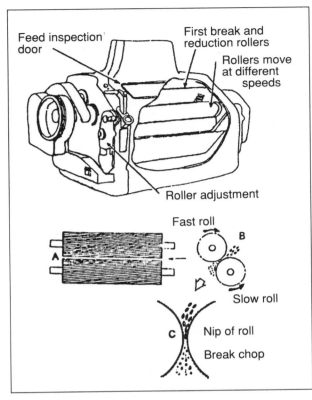

Fig. 24 *Roller mill*

ROLLER MILLING

Roller milling is used for processing either wheat or maize on a large scale. Commercial mills usually operate at between 200 to 450 tonnes every 24 hours, but smaller mills are available. The process differs from one-step mills such as stone, hammer and plate-mills, in that it is a gradual breaking down of the grain and removal of the bran. Figure 24 illustrates the basic principles of roller mills.

Operation

First the grain is cleaned and then a process known as conditioning is carried out. In wheat this toughens the bran to prevent it from breaking into very fine fragments, which would be difficult to separate from the white flour. Maize is conditioned both to toughen the bran and also to assist in the separation of the germ. The grinding is carried out by passing the grain through a series of double rolls which turn together to crush the grain. The first sets of rolls are known as break rolls which are fluted and split open the grain. The flour produced is sifted out and the residue goes to a second break roll where further grinding takes place. Once again the residue goes on through a series of similar operations until

all the endosperm has been removed from the bran. At this stage the clean endosperm is known as semolina ('half-milled') and this goes to the second series of rolls which are smooth and reduce the coarse semolina to fine flour. By further sieving and blending together flours and meals are produced for a wide variety of end uses. Roller milling is a very efficient method of manufacturing high-quality wheat flour for bread and high-quality maize meal on a large scale. The capital cost is high, good agricultural and transport infrastructure is essential, and it is not recommended unless considerable expertise is available to operate it.

By contrast, small roller-mills, with one or two sets of rolls only, are manufactured in South Africa and Europe for maize grinding. The first pair of rolls are wide-fluted and break the grain to release germ and husk. The second pair of finely fluted rolls grind the endosperm pieces to meal. A sieving system may be built into the mill. The wide-fluted rolls can be used to split legumes.

The average energy requirement for a small roller mill is about 7kWh/tonne of milled grain, which is half compared to the other mills discussed in this book. This indicates that the future in low energy milling will be in small roller mills.

Capacity

Commercial roller mills range from 50 to 1,000 tonnes a day.

Small roller mills, whether customs or commercial, usually grind from 200 to 500kg/h.

Resource requirements

Experienced and trained staff with some mechanical ability are required to operate this type of mill, in particular, to check the raw grain for metal and stones which can damage the fluted rolls, to control feed rate, check the sieves, and adjust the roll settings for different fineness of grind. As with hammermills running on customs mill basis, some financial training will be needed.

Maintenance and spares

As parts wear out they are replaced. Rolls will need recutting and sieves re-clothing. These operations are usually carried out by the manufacturer.

SOURCES OF POWER 6

In this chapter the various sources of power are described. In most African villages today grain is either hand-pounded or milled in a small commercial hammermill run from either a diesel engine or an electric motor. There are other systems available, intermediate in scale or in technology with power ranging from hand-pounding to fuel driven; for example, manual rotation, animal, water or wind power. Tables 5 and 6 outline these alternatives and indicate the scale of production for which they are suited and the kind of resources, capital, skill and raw materials required.

Table 5. Approximate sustainable power output and working periods			
SOURCE OF POWER	POWER DEVELOPED		WORKING HRS/DAY
	(W)	(HP)	
Man arms	25	0.05	5
Man legs	75	0.1	5
Donkey	200	0.3	5
Ox	500	0.7	6
Cow	400	0.6	5
Buffalo	600	0.8	7
Horse	500	0.7	5
Mule	600	0.8	8
Camel	1,200	1.6	8

Human Power and Draught Power

Figure 3 in Chapter 3 illustrates traditional pounding and saddlestone milling. The use of these tools dates back to many thousands of years. In the evolution of improved milling technique, the introduction of the lever was the next significant step which was followed by the rotary stone mill (see Figure 18). The Romans were the first to use draught animals to increase the output of mills.

The common improved hand-operated mills in use today are the simple plate mills illustrated in Figure 19. A major drawback is the relatively high investment costs in relation to output and the problems of producing fine flour. The output figures normally relate to coarse flour and they drop substantially when more fineness is required. For many women, the hand-operated plate or stone mill is only a marginal improvement compared to hand-pounding which reduces flexibility to produce desired products.

Animal traction fills the gap in terms of power available and indeed historically between manpower and machine power. Table 5 shows the power developed by many different animals and the number of hours per day they can be expected to work.

Figure 25 is a newly developed animal-powered milling technique. A horizontal arm turned by an animal around a fixed point is attached to a wheel.

Through its centre is a spindle which is attached to a wheel with a cog and chain mechanism like that of a bicycle. The other end of the chain goes round a smaller cog, through the centre of which is another spindle which drives the grinding stone itself. The wheel makes constant contact with a ring of hardened earth or concrete which the animal walks around. The mill itself rests on the horizontal arm.

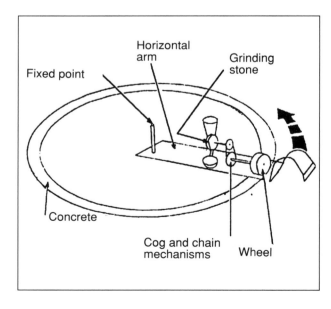

Fig. 25 *Diagrammatic representation of an animal-traction mill.*

52

TABLE 6. SOURCES OF POWER

	HAND	FOOT	ANIMAL	DIESEL	ELECTRIC MOTOR	WATER	WIND	STEAM
POWER OUTPUT, W	25	75	500	(ox)	size to suit needs	size to suit needs	design to requirement	design to requirement
THROUGHPUT, kg/h	5	15	30	180	180	50–150 (turbine)	50 (3m sail)	variable
INITIAL COSTS	low	low	fairly low	high, increasing with size	fairly high, increases with size	high	high	high
EASE OF CONSTRUCTION	easy	easy	relatively easy	easy to install	needs specialist competence	complicated, needs specialist competence	complicated, needs specialist competence	complicated
TYPE OF FUEL/COST	food	food	feed	diesel, vegetable oil, high fuel cost	electricity, cost varies, low to high	continuous running water	continuous strong wind	fossil fuel, local biomass
PORTABLE	yes	yes	no	option	option	no	no	no
OPERATION (SKILL REQUIRED)	low	low	high	high	medium to high	high	high	high
MAINTENANCE (NEED, SKILL REQUIRED & COST)	little	little	high, semi-skilled, medium cost	scheduled, medium to high skill, high cost	low, medium to high skill, medium cost	low, high skill, medium cost	high, semi-skilled, medium cost	medium/high skill, high cost
DURABILITY	fair	fair	fair, design weaknesses	low to fair, cooling problems	high, location and construction dependent	high, location and construction dependent	high	high
OTHER REQUIREMENTS	none	none	good location, trained animals	good location, regular fuel supply	good location, regular electricity supply	good location, continuous running water	good location, continuous strong wind	specialist competence and inspection
LIMITATIONS	low power	low power, social restriction	restricted power, diseases, durability	spares, service facilities, availability and cost of fuel	good electricity supply needed	distance to customers, running water	wind, no tradition in many areas	not familiar
ADVANTAGES	simplicity, locally produced	simplicity, locally produced	local power source, local manufacturing	established technique, multi-use	simple to use, established technique	local source of power, low operational costs	local source of power, low operational costs	local fuel
ECONOMY	low return	low return	profitable in remote areas	profitable with good management, back-up service and sufficient customers	profitable with good management and sufficient customers	profitable with good management and sufficient customers	low return to high investment costs	profitable only under special conditions (free fuel)

There are a number of models of animal-powered mills designed with assistance from aid agencies and they are based on similar principles. The power is transferred from the rotating wheel by chain and sprockets or through friction wheels. It is fair to say that they are still in prototype stages as only very few mills of this type have been commercially produced. Major stumbling blocks are durability of basic parts and unnecessary but costly over-engineering of other parts. However, it is not a major task to overcome these problems.

Economic surveys indicate that animal-powered mills can only compete with diesel-powered hammermills in remote areas with a costly supply of diesel. The main reason is the relatively low output in relation to rather high investment costs. Secondly, it is apparent that the type of mill selected restricts the quality of flour produced which in turn reduces the demand for its service. No doubt, there is a need to adapt the mill to the milled product desired by the customers instead of the opposite. This type of mill can be fairly cheaply constructed in developing countries by any metal worker who is familiar with welding, drilling, and construction of moving parts. In areas where animals are already in use for ploughing, the installation of an animal-powered grain mill would produce greater productivity from the livestock. One should remember that draught animals must be fed even when they are not at work, so any productivity increase is a welcome bonus. Animals are susceptible to diseases and parasites and working oxen are rare in tsetsefly infested areas as they will need prophylaxes against trypanosomiasis.

Pedal-power could be interesting as the legs are stronger than the arms. Various pedal-driven designs have been tried, and Figure 20 shows a typical arrangement where the mill is driven by a slightly modified bicycle. Social reasons have prevented the adoption of such mills. In Africa women hand-pound the grain and men ride the bicycles. Then there is a morality question concerning women using and exposing their legs. Considering the low throughput, the cost of a pedal-mill is quite high

Diesel Engines

The diesel engine is the most familiar internal combustion engine for driving machines like mills and pumps where electricity is not available. The diesel engine has a number of advantages as diesel can be replaced by locally produced vegetable oils in engines with indirect injection. Though more expensive per unit than a petrol engine, a diesel engine is cheaper to run: diesel oil is usually much cheaper than petrol and the output is some 30 per cent higher in terms of usable energy per litre of fuel. Also, diesel engines are robust (compared to petrol engines) with a long lifespan. Major maintenance may prove expensive as it requires considerable skill, though no more than for a petrol engine. Little skill is required for routine checks and servicing: for example, to top up the lubricating oil, to clean and occasionally replace the air-filter and oil-filter, and drain the water separator.

The diesel engine will continue to be an important source of power for mills for many years to come. Only 5 per cent of the hammermills in Zambia are using electricity. How will the shrinking supply of fossil fuel affect the present trend of engine development for developing countries?

Fig. 26 *Old-fashioned diesel engine*

'Old-fashioned' diesel engines

A recent interview with an Afghan miller rebuilding his old mill, which was completely destroyed by the war, provides useful information on how to select the appropriate source of power for a mill. The miller had spent 10 years as a refugee in Pakistan, partly as a miller and manager to earn enough money to buy new machinery for his mill.

The engine selected was a second-hand, 25-year-old, water-cooled, one-cylinder Ruston diesel engine (26HP). The cooling water is kept in a separate trough of the kind illustrated in Figure 26. This is still a common model in Asia. The miller had money to buy a new diesel engine but preferred the old Ruston. A summary of the reasons given for his choice are listed below:

- The engine should be water cooled. Milling dust quickly ruins air-cooled engines. The cooling of water must be done outside the building to enhance the process and prevent dust from settling into the cooling water. In addition this system provides hot water for many domestic purposes. Hardly any firewood is available.
- The engine can operate on cheap crude oil as well as on diesel. This increases both the availability and reduces the cost of fuel. In Afghanistan there is nothing like a regular supply of cheap inputs yet.
- Low speed and durable parts given long lifespan. The miller expected the old Ruston engine to outlive at least 2–3 'modern' air-cooled diesel engines.
- Good supply of cheap spares manufactured in Pakistan.
- The miller had the skill and facilities to operate and repair the engine.
- The old Ruston engine will keep his business in operation with less problems and less costs.
- Other millers had the same opinion.

Many of the arguments listed are valid also for African conditions and subsequently they cast doubts on some of the common criteria for selecting diesel engines for milling.

The 'modern' diesel engine

The common type of diesel engine used in Africa for powering a mill is portrayed in Figure 27 and has the following characteristics:

- air cooled;
- relatively high-speed engine;
- direct injection requiring high-quality diesel fuel and lubrication oil;
- fuel efficient;
- modern engine design, easy to operate and maintain, but requires special skill and facilities for repairs;
- easy to install;
- engines and spares are imported;
- constant short supply of expensive spares and skilled mechanics in rural areas.

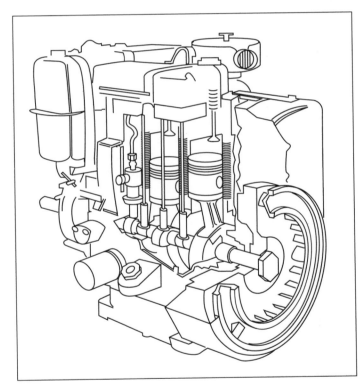

Fig. 27 *Modern air-cooled diesel engine*

The National Hammermill Project in Zambia (1,500 mills) has found that the average lifespan of an air-cooled diesel engine powering a mill is about 4–5 years totalling about 6,000hrs of operation. This is a poor performance record indicating serious weaknesses in the milling system. There are many reasons for this meagre result, one being the poor cooling caused by milling dust sticking and burning on to the cooling fins.

Shortages of spares and faulty repairs are other reasons for premature breakdowns of engines after some years of operation. The performance record from Zambia and elsewhere points towards a need to reassess the choice of diesel milling technique selected for African conditions. Major questions to look into are:

- air cooling/water cooling
- low speed/high speed
- multi-fuel use/high-quality diesel
- locally produced fuel/purchased/imported fuel
- low sophistication/high technology
- local manufacture/import
- sustainability/dependency

If the dust level in a mill cannot be reduced substantially it is obvious that water cooling is a better option compared to air cooling.

The high-speed engines are durable provided all requirements for optimum performance are fulfilled while the low-speed engines do not require the same sophistication in fuel, lubricants, filters, maintenance and repairs.

The common direct injection diesel engines will only work well on high-quality diesel. They have a mediocre performance record on alternative fuels like pure vegetable oil. Diesel engines with indirect injection, on the other hand, work equally well on pure vegetable oils. Twenty-five kilos of sunflowers contains enough oil to mill the annual flour requirement for a local household. With the appropriate type of engines, the rural villages can produce their own fuel and lubricants for their diesel engines.

Sophisticated design and production techniques exclude local manufacturing of diesel engines in most developing countries. With a minor reduction in fuel efficiency, it should be possible to commercially produce more crude types of well-tested diesel engines in a number of countries which could be operated on local fuel.

The hot-bulb engine technique

An interesting engine technique which fulfils the criteria for a sustainable power source for milling in Africa is the hot-bulb engine (HB-engine) or crude oil engine.

The HB-engine is a multi-fuel engine that operates on fuel like diesel, crude oil, vegetable oil, wood tar, fish oil, producer gas, etc. According to official tests in Sweden the fuel consumption is about 10 per cent higher compared to a modern diesel engine of equal size. Locally extracted vegetable oil can be used both as fuel and a lubricant. This means that a village in Africa can produce their own requirements of these items.

It is a water-cooled two-stroke engine as illustrated in Figure 28. The main feature is the uncooled hot bulb of the combustion chamber which ignites the injected fuel. The compression rate is only 1:6–8. The hot bulb will have to be heated before the engine will start, but after that the heat from the combustion will maintain the high temperature of the bulb. Traditionally, the bulb was heated with a kerosene pressure torch working along the same principle as a Primus heater used for cooking. It will take about five minutes to heat the bulb. An engine which is warm can be restarted without heating. The engine can then work non-stop for months provided fuel and lubricants are supplied at regular intervals.

The HB-engine or crude oil engine has few moving parts which makes it easy to manufacture

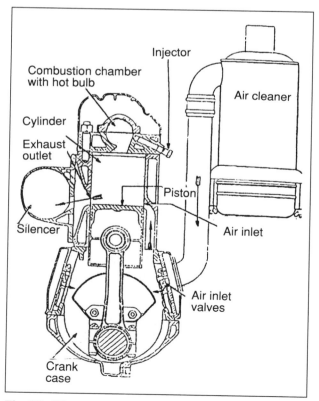

Fig. 28 *Water-cooled hot-bulb engine*

56

and repair. Most developing countries have the foundry facilities to undertake production. Village craftsmen with simple tools and spare kits can provide repair service.

These features of simplicity have a long history. The HB-engine is originally an American patent from 1887 that was acquired by a Swedish company in 1899. More than 100,000 HB-engines have been manufactured in Sweden (4–400HP) by close to a hundred small manufacturers. The engine was particularly popular in the agricultural sector and in fishing boats in the Nordic countries. Sweden was cut off from importing oil during the Second World War and had to rely on locally produced fuel. The HB-engine proved to be an excellent converter of local fuel but rural electrification gradually reduced the demand for the HB-engine type of stationary engines and the production stopped in the mid-1960s. Moreover, the good supply of cheap oil reduced the need for alternative fuel use.

Electric Motor

Most commercially available mills are supplied with electric motors as a standard fitting (Figure 29), as an alternative to a diesel engine. These

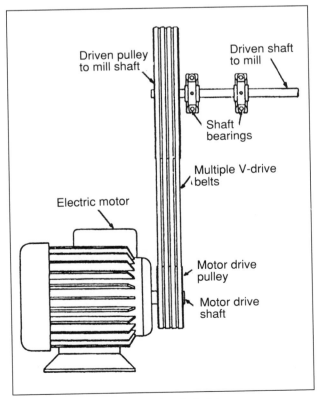

Fig. 29 *Electric motor with V-belt drive*

mills are mostly situated in towns and communities along major networks where electricity is easily available and cheap. Electric motors are simple to operate, efficient and clean and they are generally extremely reliable unless grossly overloaded. At start-up an electric motor takes a much higher current than when the mill is operating at full speed, therefore, the correct switch gear must be fitted. An electric motor is designed for frequent use and is better than a diesel engine if the mill is not used intermittently as with customs mills. Installation and repairs of electric motors must be done by licensed electricians.

Electric motors may be used to drive any kind of mill, though the most familiar are hammermills and stone mills, depending on the type of product required by the consumer. Electric motors require only a switch to start, perhaps requiring even less skill than is the case with internal combustion engines.

Water-mills

Provided that there is a regular flow of water in a river or stream or a large supply of water in a lake or dam then water-mills can be attractive solutions for the milling of grains. With controlled water in a mill race the water wheel can run on a horizontal shaft lowered to meet the water. It can be overshot, undershot, sideshot or turbine operated. The latter is the only system that one would use today. The other systems have been used with great success in the past but are relatively difficult to install and frequently require a considerable fall of water, perhaps as much as three metres, to generate the energy to drive them. A mill powered by a more efficient cross-flow turbine is shown in Figure 30. A belt drive is used to transfer the power from the turbine to the mill.

The installation costs are normally high for water mills but the operational costs are low. If the conditions are right they can compete successfully with diesel engines concerning overall costs. Except for the turbine, the structures involved must be designed to fit the topographic conditions. The design and construction must be carried out by qualified engineers while the operation can be supervised by trained operators. The turbines appropriate for milling are in the range of 3–30kW and they are regulated (power and speed) through the adjustment handle shown in Figure 30.

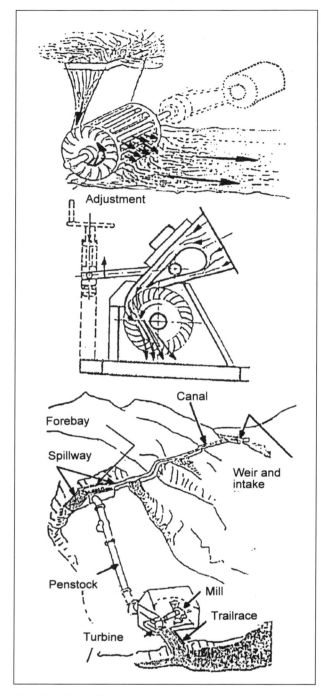

Fig. 30 *Cross-flow turbine*

A turbine can be used for powering more than one machine.

The cross-flow turbine can be manufactured by local metal workshops. A disadvantage is that the mill can only be installed where there is sufficient quantity and slope of flowing water. This restricts the possibilities of erecting the mill in good business areas with enough customers. Any survey on appropriateness of a site for a water-mill must include relevant market studies and the potential income generated from milling.

Anybody interested in water-mills is recommended to study *Micro-Hydro Power* published by IT Publications. The book will also give relevnt information on organizations and references covering the subject.

Locally manufactured cross-flow turbine-driven mills are popular in Nepal. They also have proven successful in Ethiopia where about 20 plants are in operation today. It appears that densely populated hilly areas with streams flowing during most parts of the year are the most suitable areas for water-mills.

Wind Power

Though wind power appears attractive, it is in fact not a viable alternative in most situations. The wind must blow steadily and frequently and the windspeed should be in the narrow range 3–15m/s (10–50km/h). The traditional sail and wood design of windmill, though it appears simple, is very difficult to construct and drives large stones through complex gears. Modern windmills of the kind used for water pumps may appear suitable but the disadvantage of intermittent wind supply, which does not affect the collection of water into a container, makes them unsuitable for milling. Wind generators charging batteries to drive a DC-motor have not proved to be successful for milling purposes.

Battery-powered Mills

A mill can be driven by electricity from a battery charged by solar cells or a wind-driven generator. However, grinding grain to flour is a relatively power-demanding operation which sets limits to the size of a profitable battery-powered milling plant for village use. The technique applied is well tested but it is still at prototype stage for grain milling as little work has been devoted to reduce the energy requirement of the grinding operation which is a key factor with a limited power source like a battery. The energy requirement varies between 0.7–5.0kWh per 100kg ground product for unfermented grain. The lower figure comes from roller mills while hammer/plate/stone mills will require at least double this figure. Type of grinding, condition of mill, fineness and feeding

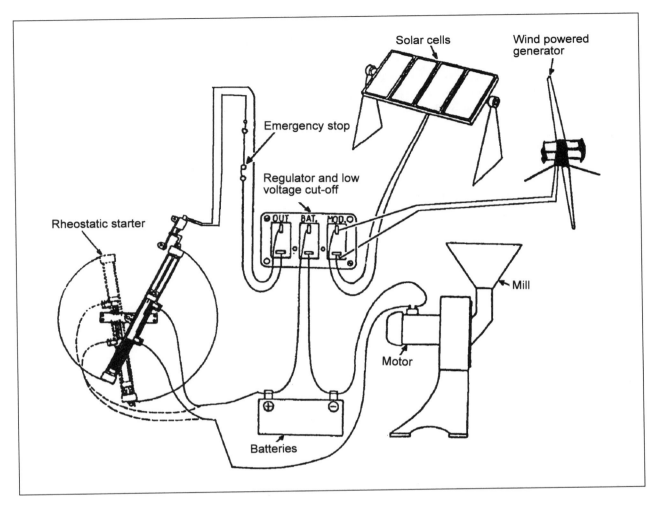

Illustration 1 *Solar and wind-driven layout*

rate are main determining capacity factors. Dry milling of fermented grain/grits is an old traditional system for reducing power requirement for milling which ought to be tried as a first step in increasing efficiency. Illustration 1 above describes the principles of a battery-powered milling system.

The simple and sometimes crude design of electric and diesel mills manufactured in Africa are copies of old mill designs and have hardly undergone any extensive research on how to minimize power requirement. A common habit is to increase engine size in order to obtain better performance of simple mills which is a rather wasteful mechanization trend as a good mill will cost much less than a diesel engine.

A study of testing results of mills in Sweden shows large variations in energy requirement per quantity milled product. The simple types of mills promoted in many developing countries belong to the least-efficient models. This indicates a potential to improve mill performance and select the most efficient type of mill in relation to energy requirement per kilogram milled product. The small roller mills appear to be the most realistic alternative as they can double the quantity milled per battery compared to hammer/plate/stone mills.

A set of 60 photovoltaic modules of 43Wp each in combination with 30 automotive batteries (90Ah and 12V) is capable of grinding 200kg grain per day according to field experience. This is too low a capacity for a customs mill. The total milling capacity needs to be substantially improved to be a profitable alternative for an investment of US$10,000. A diesel-powered customs mill of equal cost needs to grind close to 1,000kg per day to achieve break-even cost coverage under Zambian conditions. Even if the cost for diesel (one-third of cost) is excluded in the comparison, the battery-powered alternative needs substantial improvements in capacity to be a realistic alternative for rural areas in Africa.

The same arguments are valid for charging batteries with wind-powered generators. A Danish test on direct connection between mill motor and wind generator was able to achieve a milling rate of 20–30kg grain per hour with an average windspeed between 5–7m/s.

Another weakness with battery-powered mills is the need for specifically designed motors, switches and control boxes, which are different from the common electric standards. The rheostatic starter mentioned in Illustration 1 protects the motor from overheating during the starting phase and when the batteries begin to be discharged and fail to turn the motor.

Steam Engines

The external combustion or steam engine has been used for two centuries to drive machinery. It is relatively easy to maintain and the source of heat for generating the steam is obtained by burning biomass such as wood, husks, coconut shell, coal or oil. The steam generated is then used to drive the steam engine. The disadvantages are that a steam engine must have an external source of fuel and must have clean, purified water for the generation of the steam. Steam engines are relatively expensive in that they require a furnace, a boiler to generate the steam and a relatively complex mechanism to convert the energy of the steam into rotating mechanical energy. Thus these engines are very expensive compared to diesel and/or electric engines and they have only received widespread use in recent years in countries where a biomass residue, such as rice husk, is available at negligible cost. As with the diesel engine a steam engine has the advantage that it can be used to drive an electricity generator. For very small-scale milling alone, the steam engine is not a practical solution. For large-scale milling, particularly of rice, it may be the case, as the fuel is virtually free of charge. It is important to note that boilers for steam engines are potentially very dangerous pressure vessels and require regular inspection. In a village environment with unskilled labour, a steam engine is not a viable alternative.

Coupling Arrangements

Coupling arrangements are required to engage or disengage the source of power to the mill and to ensure the desired revolution of the machinery. The principal types are shown in Figures 31–36.

Mills are driven from engines/motors directly or by belt drives which may be V-belts or flat-belts. It is advantageous with some types of mills to have a clutch so that the mill is not being driven when not grinding, or, alternatively, a flat-belt with shifting fork to disengage power. Diesel engines use relatively little fuel when disconnected from the mill (idling) but the wear on the engine is then higher. The rotating shafts and belts should be guarded to protect operating staff as they can cause severe injuries.

Direct drive

Some milling equipment like hammermills can be permanently engaged to an engine if the engine can match the required periphery speed of the hammers. Prerequisites are that the mill should be able to idle if necessary and be difficult to overload. This means a simple and durable direct drive which is pictured in Figure 31. This arrangement will allow for easy transporting of the mill between different milling plants.

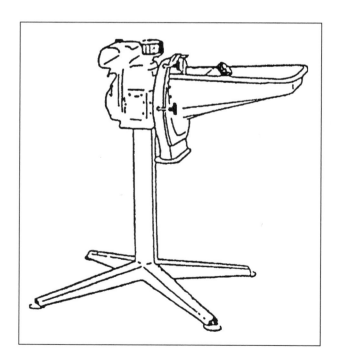

Fig. 31 *Direct drive*

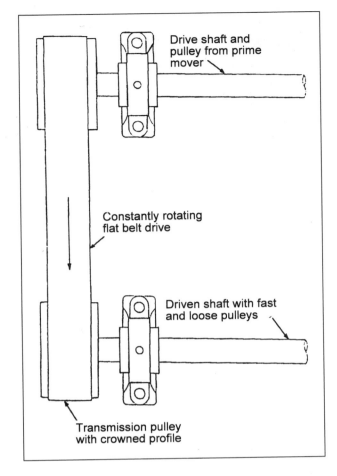

Fig. 32 *Flat belt drive*

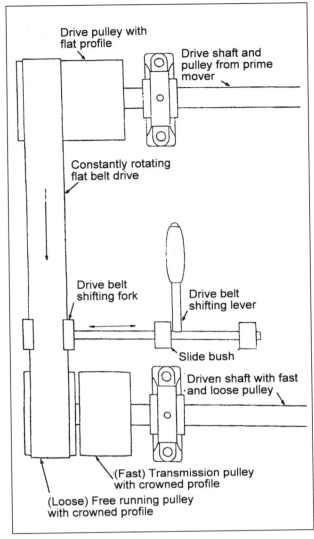

Fig. 33 *Flat belt coupling arrangement*

Flat belt drive

Figure 32 shows the drive from a prime mover (engine, motor) being transmitted to a pulley fixed to the shaft through a flat belt. Wax is used to increase friction between belt and pulleys to avoid slippage.

Drive is transmitted to the loose pulley in Figure 33 via the flat belt. As the name suggests the loose pulley is free running on the driven shaft. The driven shaft will not turn until the belt is shifted with the belt shifting fork on to the fast pulley which is fixed to the driven shaft (see Figure 33).

To prevent the belt from slipping off the pulleys, both loose and fast pulleys have a crowned profile, based on the principle that flat belts will always ride on the largest diameter, the belt will therefore run in the centre of the pulley. The drive pulley is plain, as long as the loose/fast pulley assembly.

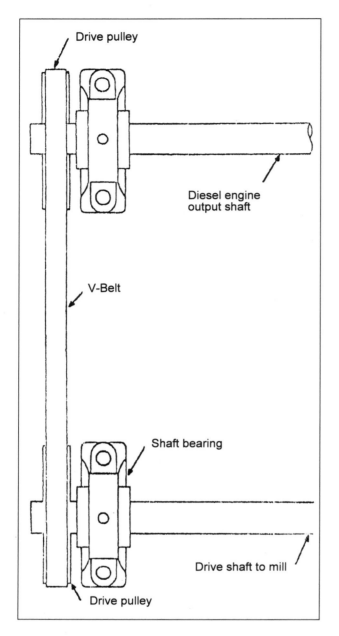

Fig. 34 *V-belt drive*

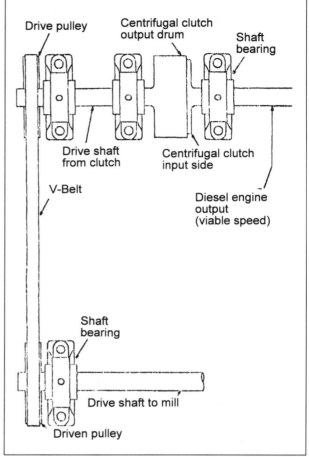

Fig. 35 *V-belt drive with centrifugal clutch*

Centrifugal clutch coupling

Figure 35 shows the drive from a diesel engine output shaft. The diesel engine is set to run at two speeds, idle, and normal running speed. The output shaft is fixed to the input side of the centrifugal clutch. The input side of the centrifugal clutch is fitted with friction pads, similar to motor vehicle brake shoes. When the engine is idling the friction pads are running clear of the clutch output drum. When the engine approaches operating speed the pads are thrown out to make contact with the output side of the clutch. The clutch is able to transmit full power when the engine is running at its correct speed. Drive is transmitted into the drive pulley, and driven pulley via a V-belt. As the pulleys are fixed to their shaft, drive is transmitted to the mill.

V-belt drive

Figure 34 shows a standard V-belt arrangement. Drive is transmitted to the driven pulley via a matched set of V-belts capable of transmitting full power. The driven pulley is fixed to the shaft and drive is transmitted to the mill. No clutch is used in this arrangement as the mill can be stopped by simply switching off the motor in the case of electric motors. This type of coupling arrangement is also common for diesel-driven hammermills.

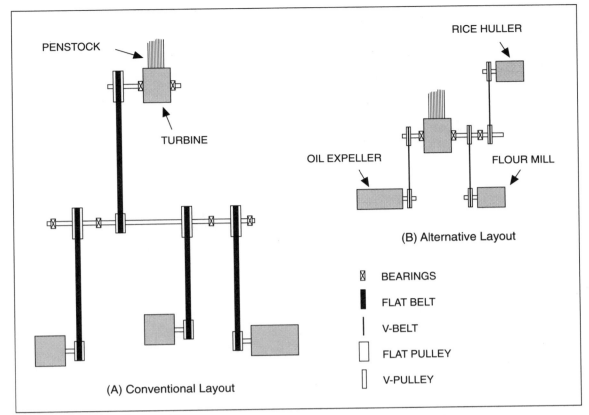

Fig. 36 *Belt drives from turbine*

Multiple drives

It is common to have one power source to drive a
number of machines through systems of belt drives.
Figure 36 shows the arrangements for a turbine to
drive three different machines.

CASE STUDY 3

The main purpose of this case study is to describe the activities involved in basic milling operations of diesel-driven hammermills in Zambia through an Operator's Manual. The reason for selecting a diesel-driven hammermill as a case study is because it is the most common type of mill in Africa. Similar activities are involved for the other milling machines described in the book. The Operator's Manual is also a practical training manual for trainees working in a mill for at least two weeks under the supervision of a trained operator (learn by doing). More detailed pictures and explanations are required for a 'teach yourself manual'. This case study is highly relevant for developing countries.

Introduction

The Village Industrial Service (VIS) in Zambia is participating in the National Hammermill Programme and assisting approximately 150 mill entrepreneurs in rural areas. In addition to their own programme, they also assist in training hammermill operators for other programmes. The Operator's Manual is based on their training experience of operators and most of the drawings in this case study originate from their training material which is made to cover a specific mill and engine model. The drawings are made by local artists from photographs and the manufacturers manuals. With this technique, the basic requirement for making an Operator's Manual is a camera, typewriter, manufacturers' manual and a copy machine. The advantage of using many drawings is that people with a poor educational background can more easily understand the manual.

Added to the text is the experience of the Small Industrial Development Organization (SIDO), Zambian Cooperative Federation (ZCF), Technical Development and Advisory Unit (TDAU) of the University of Zambia and the Agricultural Engineering Section (AES) of the Ministry of Agriculture.

The mill is a Saro mill of the 'Hippo type' which is a simple hammermill model which consists of an axle with hammers and a screen underneath. The milled product passes through the mill and is collected in a bag under the mill. A major drawback with this design is dust.

The engine is a two-cylinder, four-stroke and air-cooled direct injected Kirloskar diesel engine rating 15hp at 2,000rpm. The mill and engine are mounted on a frame which is bolted to a concrete foundation (Case 3.1). The power is transmitted through a V-belt drive of 3–4 belts. A blanket is placed as a divider between the engine and the mill to protect the former from dust. The walls have large openings for ventilation and light.

The Hammermill Operation

Daily pre-operation tasks

It is important to ensure that the engine is ready for the daily milling work before starting the engine. The daily routine involves checking and filling of oil and fuel.

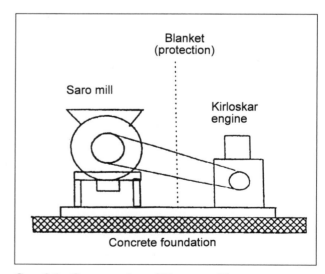

Case 3.1 *Cross-section of Hammermill*

Checking and filling of oil (Case 3.2)

1. Clean the area around the dip-stick and the oil cap with a cloth.

2. Unscrew the dip-stick and pull it out.

3. Wipe off the oil with a clean cloth.

4. Replace dip-stick.

5. Pull out dip-stick again and check oil level.

6. If oil level is below minimum mark, remove oil cap and add some oil. Oil level should be between minimum and maximum marks. Repeat procedure to ensure correct level of oil.

7. Replace oil cap and dip-stick and ensure that they are properly tightened.

8. Record amount of oil added.

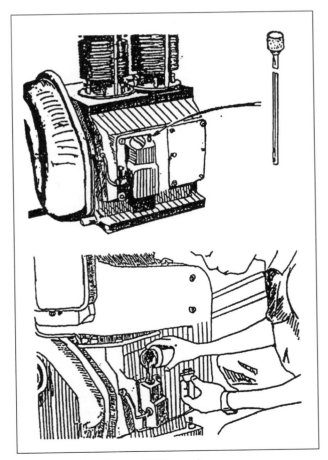

Case 3.2 *Checking and filling oil*

Case 3.3 *Checking, filling and recording*

Checking, filling and recording of fuel (Case 3.3)

1. Clean fuel tank with a cloth.

2. Unscrew filler cap of fuel tank and remove strainer.

3. Check fuel level and replace strainer. Never fill diesel without the strainer in place!

4. Refill as necessary using a funnel and a graduated container.

5. Replace filler cap and ensure that it is properly secured. A lost filler cap must be replaced as soon as possible. Never use a dirty cloth as a filler cap replacement!

6. Record in the daily record ledger any amount of fuel and oil added.

4. Crank the engine. The faster, the easier the engine will start.

5. Disengage one decompressor lever by turning it down and keep on cranking.

6. When the engine starts to fire, remove handle.

7. Push the other decompressor level down. When both cylinders are properly firing, the engine picks up speed and sounds normal, then start to mill.

8. Occasionally the engine refuses to start and no smoke comes out of the exhaust pipe. A common fault is air in the fuel system which prevents the fuel from being injected. If this happens the fuel system must be 'bled':

Engine operation tasks

Starting and stopping of engine (Case 3.4)

STARTING
Careless use of the starting handle can easily cause damage to the camshaft extension shaft, the clutch pin and the handle and necessitate expensive repairs. Ensure that the handle is properly engaged before attempting to crank the engine. If free play is developed, replace worn parts.

Never grip with the thumb around the starting handle. The thumb should point in the same direction as the fingers on your right hand while cranking as an unfortunate back-stroke can break your thumb!

1. Engage the starting handle properly to the camshaft extension shaft.

2. Release overload stopper which is fitted on the fuel pump of some models.

3. Engage the decompressor levers by turning them upwards.

Bleeding the fuel system

- Open air-ventilation screw on the fuel
- filter housing. Diesel and air in the fuel system from the tank to the filter will flow out. When no air bubbles are visible, tighten the air-ventilation screw;
- Continue with opening the next air-ventilation screw(s) on the fuel pump(s) and diesel and air bubbles will start to flow. When air bubbles disappear, tighten the air-ventilation screw;
- Try to start again repeating the above procedure;
- If the engine continues to fail there may be air bubbles left in the high pressure pipe between the fuel pump(s) and the injector(s). Carefully loosen one of the pipe nuts slightly where the high pressure pipe enters the injectors and crank the engine. Leaking diesel will appear forcing the air out. When no air bubbles are visible, tighten the pipe nut carefully and repeat the procedure on the next injector. Ensure that you have the right spanners and screwdrivers for this operation.
- Repeat the starting procedure.

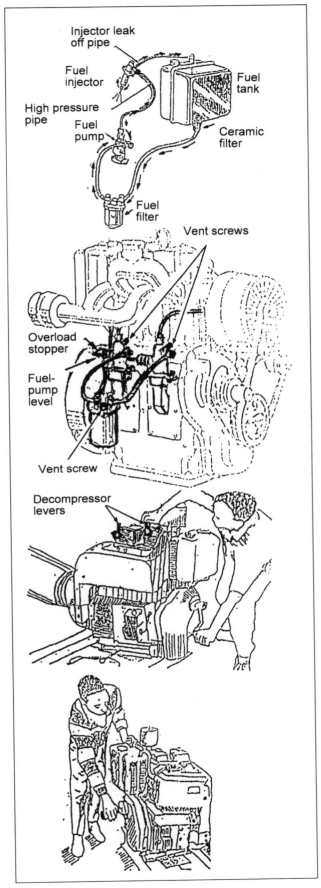

Case 3.4 *Starting and stopping the engine*

1. Press fuel-pump lever towards the pump element and hold it there until engine stops.

2. Never use the decompressor levers to stop the engine!

Milling tasks

Milling and recording grain (Case 3.5)

Never let uncleaned grain into the mill. Soil and stones quickly ruin the mill and cause costly repairs. Many millers require that the grain must be properly winnowed at the mill site and be inspected before being milled. This ought to be a standard routine in any customs mill.

Most customers want their own grain back after milling. Normally they bring with them 1-2 tins (15–30kg). This means that the mill must be emptied between each customer.

The Saro mill produces a lot of dust which is unhealthy to inhale. Always wear a cloth around the mouth and nose when grinding. One way to reduce dust is tightly and permanently to fit a bottomless bag or stocking to the outlet opening and insert into the bag to be filled with milled products.

Never stop the engine with grain in the mill!

Milling charges are either for the number of tins to be milled or the number of tins of flour received from the milling process. The former is to be preferred as insect-damaged grain will produce less flour. Also business is smoother if charging is done beforehand.

1. Ensure that the grain is properly cleaned. Refuse uncleaned grain!

2. Fill the grain into tins. Take note of the numbers and collect milling charge.
   Record tins and money received for every customer in a notebook to be added up for the daily records form.

3. Check that the mill pan is empty before serving the next customer.

4.　Attach bag to the mill hooks.

5.　Put on cloth covering mouth and nose.

6.　Pour grain into mill pan.

7.　Open feed control gate slowly. If black smoke comes out of the engine, it is a sign of overloading. Reduce the feed control gate opening until the exhaust smoke is white. An even feeding will cause a more uniform flour quality.

8.　When the customer's grain is milled, close the feed control gate and remove the bag with flour from the mill hooks.

9.　Continue with the next customer.

Post-operation tasks

Cleaning hammermill and shelter daily (Case 3.6)

Due to the dusty conditions, it is vital to have a daily cleaning routine. This will reduce the fire hazard, cooling problems of the engine, rat and bird infestations and improve the health of operators as old and mouldy dust can be extremely dangerous to inhale. A clean and appealing mill will attract customers while a filthy place will scare many off.

Case 3.5　*Milling and recording*

Case 3.6　*Cleaning of hammermill shelter*

One can clean the mill after the daily milling work is completed but one cannot clean the engine before it is cold. As milling normally takes place during the later parts of the day it is common to clean the following day before the customers start to line up.

1.   Sweep the shelter with a broom.

2.   Dust the engine with a cloth. The engine should look like a new and shiny engine every day! It is much easier to wipe off fresh dust than old.

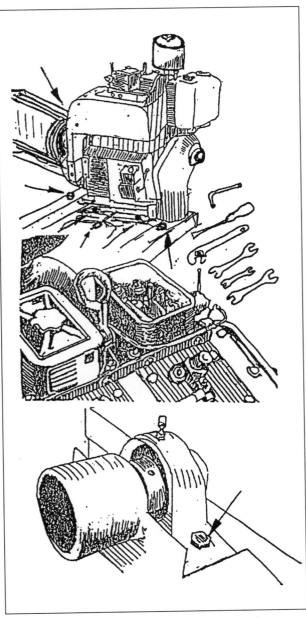

Case 3.7   *Check and tighten selected bolts and nuts*

Hammermill Maintenance

The milling operation causes vibrations and wear. Loose bolts and nuts can cause premature and unnecessary breakdowns as dirt and will reduce the lifespan of an engine. It is important to have a set of proper tools. The supplier will inform about the right types and sizes required. Avoid using adjustable spanners as they tend to damage bolts and nuts more easily. In addition, wrong tools slip readily to cause personal injury. **As a tool set costs a fraction of a mill, there is no excuse for not having proper tools!**

Weekly maintenance tasks

The weekly maintenance should be carried out on the same day every week. Monday morning is a common time for this job.

Check and tighten selected bolts (Case 3.7)

If bolts or nuts are missing, replace them. If loose, tighten them.

1.   Check engine mounting bolts.

2.   Check foundation bolts.

3.   Check V-belt guard bolts.

4.   Check mill bearing bolts.

5.   Open rocket box covers:

• check peg screws with flat screw driver;
• check grub screws with Allen keys.

Case 3.8 *Check and tighten beater bolts*

Check and tighten beater bolts (Case 3.8)

A nut or hammer falling off will seriously damage the screen and render the mill out of order.

1. Remove the eye bolt from mill cover and open it.

2. Mark the first bolt with charcoal.

3. Check and tighten any loose beater bolts

    and nuts using two 17mm spanners. That means using one spanner for holding the bolt and the other for tightening the nut.

4. Return the mill cover and replace the eye bolt. Do not start the engine when beaters are uncovered!

Cleaning of air-filter elements (Case 3.9)

Dust which is allowed to enter the engine will increase wear substantially. Therefore, it is essential to remove dust from the inlet. This function is performed by the air cleaner with filters. The engine can be equipped with dry or wet air filters. The first one is made from specific types of papers while the other type has a wire strainer in oil bath. Never operate an engine without an air filter!

(a) Dry type

1. Remove the wing nut and the flat washer on top of the air filter.

2. Remove the filter cover bowl.

3. Remove the air cleaner element.

4. Tap the air filter element carefully against piece of wood until all dust (meal powder)

is removed. If there is a hole in the filter it must be replaced immediately.

5. Clean the filter cover bowl with a dry cloth.

6. Refit the air filter element and bowl, flat washer and lastly the wing nut.

(b) Oil bath type

1. Remove the wing nut, washer and bowl.

2. Loosen the clip with screw driver and remove oil bath container.

3. Take out the wire strainer and empty the dirty oil.

4. Clean all parts in kerosene or diesel and wipe dry.

5. Reassemble oil bath container.

6. Fill new oil up to the marking in the oil bath.

7. Reassemble remaining parts.

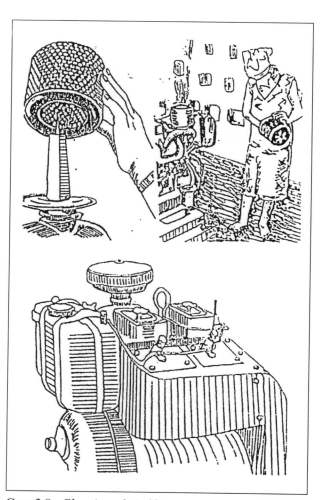

Case 3.9 *Cleaning of air filter*

Cleaning shelter and divider (Case 3.10)

Old and mouldy dust gives the flour a bad taste and speeds up its degradation. Various kinds of storage insects, which infest both grain and flower, thrive in the flour dust. Cover your mouth and nose while cleaning inside the milling shed.

1. Remove divider, take it outside and shake it clean.

2. Clean dust from the floor, the ceiling and the walls.

3. Clean the surrounding area where women are waiting and winnowing their grain.

4. Wipe engine, mill frame and mill. Make sure that there is no old flour left where the bag is attached!

Case 3.10 *Cleaning shelter and divider (blanket)*

Monthly maintenance tasks

On each fourth Monday it is time to perform the regular monthly maintenance tasks. The cooling of the engine will be seriously reduced with dust sticking to the cooling fins. If the dust is properly controlled, it might be sufficient to make a monthly check of the cooling fins. However, if it is found that dust is carbonized on the fins and must be scraped off, the cleaning must be performed more often. This can be a serious sign of over-heating which can quickly ruin the engine!

As it is a time-consuming job to clean the cooling fins, it is always better to try keeping dust away from the mill. Other types of mills have cyclones which reduce dust substantially.

Cleaning of cooling fins (Case 3.11)

In addition to spanners, a 30cm long piece of strong wire, steel brush, screwdriver, cloth and kerosene are needed.

1. Disconnect leak-off pipe using a 10mm spanner.

2. Loosen the two bolts on the fuel tank adapter plate.

3. Remove one bolt and tilt the fuel tank.

4. Remove side and main cover bolts using spanners, and lift off main cover.

5. Clean the cylinder liner fins with a brush (preferably a steel brush).

6. Clean the cylinder head fins and nozzle socket fins with the wire.

7. When no deposits are left, blow and brush all dirt away from the fins.

8. Wipe off the cowling and all fins with a cloth dipped in a little kerosene or diesel.

9. Reassemble all parts starting with the cowlings.

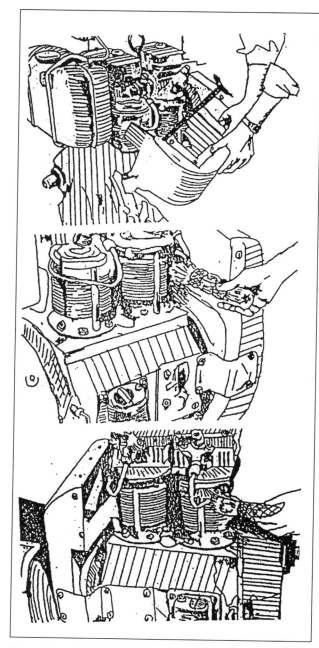

Case 3.11 *Cleaning of fins*

A standard recommendation on tension is that you should be able to push the belt down with your thumb the same distance as the thickness of the belt.

1.   Loosen engine mounting bolts.

2.   Adjust belt tension by tightening the adjusting bolt.

3.   Check engine alignment carefully and adjust if needed.

4.   Check tension of each belt. It should be the same for all belts.

5.   Tighten engine mounting bolts.

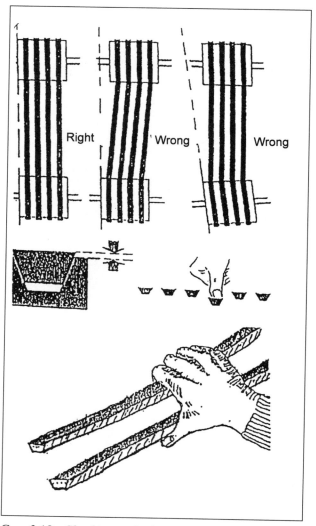

Case 3.12 *Checking and adjusting of V-belts*

Checking and adjusting V-belts (Case 3.12)

Multiple V-belt drive belts should be of equal length and tension. If they vary, only the shorter ones will pull properly. This means that when changing belts one should change the whole set. The right tension and alignment is critical for long V-belt service. It is of the utmost importance that the drive shafts are parallel and that the belt is not touching the bottom of the grooves in the pulley.

Maintenance tasks after 300hrs of operation

The engine and mill bearings need regular maintenance after 300hrs of work. By adding up the daily and weekly records, one will know when it is time for this service.

Greasing of mill bearings (Case 3.13)

1. Remove clamp bolt and open mill cover.

2. Remove grease nipple covers on both bearings.

3. Position the grease gun on the nipple and begin to pump grease into the bearing. Stop when grease starts to come out from the bearing sides. Grease both bearings.

4. Replace the nipple covers and wipe the bearings free from any surplus grease with a cloth.

5. Replace mill cover and tighten clamp bolt.

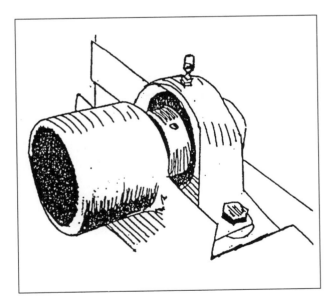

Case 3.13 *Greasing of mill bearings*

Changing of oil and replacing oil filter (Case 3.14)

Ensure that the engine is clean and that you have a trough which can hold about 10 litres of oil which fits under the oil drain plug. Run the engine for 10 minutes to warm up the oil to allow for a more complete drainage. Never refill with used engine oil even if it looks clean.

1. Remove drain plug with 17mm and 19mm spanners and empty the engine waste oil into a trough beneath the engine.

2. Remove the lubrication oil pipe from the crank case. Leave it loosely connected to the oil-filter casing.

3. Remove the casing and inspection cover.

4. Remove used oil filter and clean the filter casing with kerosene or diesel.

5. Refit drain plug and tighten properly..

6. Fit new oil-filter element with centre bolt and washer into the filter casing.

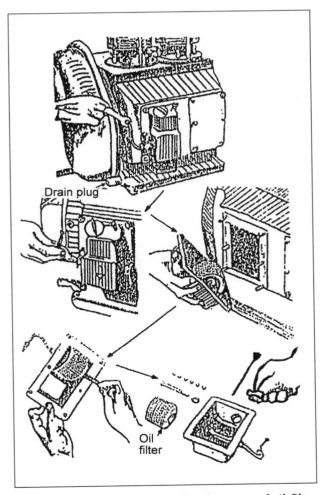

Case 3.14 *Changing of oil and replacement of oil-filter element*

7.   Refit the oil-filter casing and the lubrication oil pipe to the crankcase and tighten properly.

8.   Refill new clean oil to correct level (seven litres of SAE 30 or SAE 40).

9.   Remove the waste oil and any spillage and clean up properly. Do not throw the waste oil away as small quantities can contaminate drinking water!

Changing of fuel-filter element and cleaning of ceramic filter (Case 3.15)

Dirt can easily block the fuel system and therefore, it is of great importance to adhere to strict cleanliness of all parts involved in changing the fuel-filter element. Ensure that the fuel system is properly cleaned with a clean cloth before changing the filter.

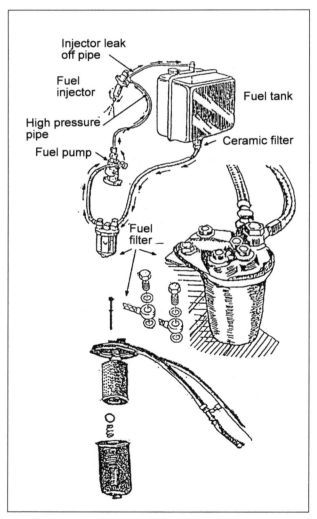

Case 3.15 *Changing of fuel-filter element*

Note that Case 3.15 shows the fuel system of a one-cylinder engine. Another fuel pump and pipes are added for the two-cylinder model. Change filter when the tank is almost empty to avoid wasting fuel.

1.   Put a basin big enough to hold the remaining quantity of fuel left in the tank below the engine where the fuel system is fitted.

2.   Remove the banjo bolt holding the fuel line from the tank to the filter and empty the tank into the usual fuel container.

3.   Unscrew the nut holding the fuel line from the tank and clean the ceramic filter in clean diesel. A blocked ceramic filter is a common reason for poor engine performance.

4.   Refit the ceramic filter and fuel line to the tank.

5.   Remove the centre bolt of the fuel-filter casing and empty the casing of fuel.

6.   Remove filter, washer and spring and discard the filter.

7.   Clean the casing from eventual deposits and replace spring first and then washer and new fuel filter.

8.   Make sure that the sealing ring is in position and reassemble the casing with filter. Tighten the centre bolt.

9.   Refit the fuel lines and fill up one quarter tank. Check for leakages. If none, fill up tank.

10.   Bleed the fuel system as described in Case 3.4.

Changing of air-filter element (dry type) (Case 3.16)

One can only clean a dry type of filter a couple of times and then it needs to be replaced (see Case 3.9).

1.   Remove the wing nut and the flat washer on top of the air filter.

Case 3.16 *Checking of air-filter element*

Case 3.17 *Removing soot from exhaust silencer*

2. Remove the filter cover bowl.

3. Remove the air-cleaner element and discard it.

4. Clean the filter cover bowl with a dry cloth.

5. Refit the air-filter element and bowl, flat washer and lastly the wing nut.

Removing soot from exhaust silencer (Case 3.17)

A silencer filled with soot will reduce the exhaust flow and the power of the engine. Therefore it is necessary to remove soot regularly from the silencer.

1. Remove exhaust silencer from the exhaust pipe (extension pipe) by unscrewing it.

2. Tap the silencer against a wooden block outside the mill until all soot is removed.

3. Refit the exhaust silencer to the exhaust pipe.

Hammermill Repairs

A skilled mechanic is required for making major repairs of the engine and mill. However, there are some repairs which can be made by the operators.

Operation repair tasks

Removing and replacing of damaged screen (Case 3.18)

There is a normal wear of screens and it increases with dust in the grain until it breaks. Damage to a screen is also caused by small stones or metal pieces.

1. Remove the eye bolt from mill cover and open it.

2. Remove the damaged screen.

3. Fit a new screen.

4. Return the mill cover and replace the eye bolt.

Case 3.18 *Removing and replacing screen*

Turning or replacing beaters/hammers (Case 3.19)

The output of the mill drops with worn-out beaters. The quality of the steel in the beaters determines when it is time to turn or replace them. A set of hammers will last between 600–1,200 hours. This means turning after 300–600 hours. It is bad business to operate a mill with worn-out hammers as it will increase fuel consumption and wear per kilogram of milled product.

1. Remove the eye bolt from the mill cover and open it.

2. Unscrew the beaters using two 17mm spanners.

3. Inspect bolts and nuts and replace them if worn out. Use the same size as the previous bolts.

Case 3.19 *Turning/replacing of beaters*

4. Refit the turned beaters/hammers into the same place to avoid imbalance and tighten them properly.

5. Replace all beaters at the same time and tighten them properly. Different weights of beaters will cause imbalance and vibrations which will ruin the bearings and the mill.

6. Return the mill cover and replace the eye bolt. Do not start the engine when beaters are uncovered!

Changing of damaged V-belts (Case 3.20)

Refer to discussion in Case 3.12. If properly fitted and adjusted a set of V-belts should last for at least 1,200 hours. However, experience shows that they will only last for one-third or half that time. This indicates poor V-belt management.

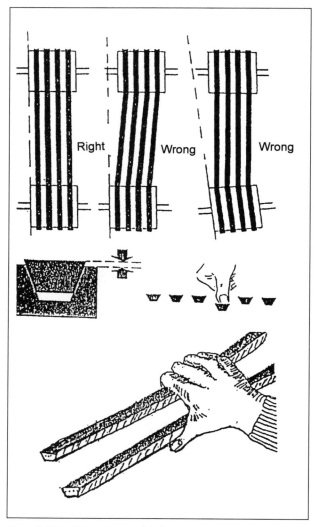

Case 3.20 *Changing and adjusting of V-belts*

1.   Loosen engine mounting bolts and adjusting bolts.

2.   Push engine towards the mill.

3.   Remove V-belts.

4.   Replace with new set of belts.

5.   Push engine away from the mill to tighten the belts.

6.   Check engine alignment carefully and adjust by tightening the adjusting bolts.

7.   Check tension of each belt. It should be the same for all belts.

8.   Tighten engine mounting bolts.

9.   Re-check the alignment and readjust if there is even a slight distortion of align ment. The extra hours spent on alignin V-belts properly is time well used.

Major repairs

When and how to get outside repair service (Case 3.21)

The need for major repairs by skilled mechanics starts after approximately two years of operating the mill. Therefore it is important for the operator to know when it is time to ask for outside assistance. Many of the engines and mills which are out of order could have been brought back in operation within a short period if they had been repaired by skilled mechanics. The problems with faulty repairs in combination with poor spare services are major reasons for premature breakdowns in rural areas.

The cost of a breakdown of a mill includes both spares and repairs but also the cost of the standstill through lost business and fixed costs which soon can exceed the total cost of the repair. Therefore, it is important to reduce the time of standstill and look for quality repairs to limit future breakdowns. It is always bad business to use unskilled mechanics even if they are cheap as they can cause future costs. It should always be the total cost which is taken into consideration.

Serious dealers, promoters and/or financiers of milling must ensure that the mill owners are aware of how and where to get assistance with major repairs and spares. There should be a list visible on the wall in every mill stating how to get adequate repair service.

Everything turns hectic when the mill suddenly stops and customers are queuing outside. Before rushing away it is essential that the owner and operator, if different, have discussed the faults properly and analysed both problems and actions to be taken. Keep `spanner boys' away from the engine and mill and consult only experienced mechanics.

1.   Identify the problems of the engine/mill by using the maintenance manual provided by the dealers.

2.   Be sure of the problem or be able to repair/ narrate it properly.

3.   The first thing to do when looking for spares is to identify serial/model numbers of both engine and mill.

4.   Use part lists to identify the part(s) required and record the spare part numbers. Always quote serial/model numbers of both mill and engine.

5.   Bring the broken parts with you if possible together with serial numbers of mill and engine when buying replacement spares.

6.   If you foresee that obtaining spares will involve substantial travelling, consult the best local experts and ask for assistance to ensure proper problem identification.

7.   Only use qualified mechanics with a proven track record. Ask for trades certificates.

Management

Record keeping (Case 3.22)

Good management is a precondition for successful milling operations and good records form the basis for good management. A mill without records is a poorly managed mill, likely to be unprofitable and soon cease operating.

An important responsibility of the operator in a customs mill is to fill in basic records of working hours, tins milled, money received, fuel and oil used and type of service and repairs made.

When commercial milling is included it is important to keep records of raw materials, quantity milled and sold (both main products and by-products).

The operator will have to work with two types of records, daily records and weekly records. They must be properly filled in after each day of work. A standard rule is that the operator is not allowed to leave work without having completed the records properly.

In addition to the daily and weekly records there is also a need to keep a cash book, to make projections and profitability analyses and to make cash-flow analysis. This task is beyond the normal operator's responsibility and is therefore not included in this case study.

Daily Records Form (Case 3.23)

The figures presented in the daily records is an extract from a diesel-driven customs hammermill and they are filled in by the operator. The usual by-product in milling is bran. The format used covers both customs milling and small-scale commercial milling. The currency used in Zambia is the Kwacha. In order to save writing space, use only the initials when recording working hours for workers including the operator.

The daily records are summarized at the end of the week and added to the weekly records.

The Weekly Records Form (Case 3.24)

The Weekly Records Form is a cumulative record. This means that the new weekly record will be added to the previous weeks. Through this system it is possible to know when it is time for the 300-hours service. This data will form the basis for calculations of milling charges and the annual result of the mill. Other information like repairs are recorded under remarks.

CASE 3.23

DAILY RECORDS FORM

OPERATOR'S FORM

YEAR_____MONTH_____TYPE OF PRODUCTION_____OPERATOR_____

WEEK_____ -- _____NO. _____NAME OF PLANT_____

DAY HRS/ DAY	PRODUCTION		SALE		REVENUE	INPUTS				
	MAIN PRODUCT Tin (15 kg)	BY- PRODUCT Tin (7.5kg)	MAIN PRODUCT Bags ( )	BY- PRODUCT Bags ( )	CASH RECEIVED Kwacha	RAW MATERIA L Bags ( )	FUEL Litres	LABOUR Initials and hrs worked /day	OTHER INPUTS, MAINTENANCE, REMARKS
MON 6 HRS	60				2,280		18	BB 7 CC 3	
TUE 5 HRS	50				1,900		15	BB6	
WED 7 HRS	70				2,660		21	BB 8 CC 3	220 l diesel drum
THU 4 HRS	40				1,520		12	BB 5	
FRI 7 HRS	70				2,660		18	BB 8 CC 2	1/2 oil added Tighten fan belt
SAT 7 HRS	70				2,660		18	BB 8 CC 3	
SUN 0 HRS	0				0		0		
TOTAL 36 HRS/WEEK	360				13,680		102	BB 42 CC 11	

CASE 3.24

WEEKLY RECORDS FORM

OPERATOR'S FORM

YEAR _____ TYPE OF PRODUCTION _____ OPERATOR _____

MONTH _____ NAME OF PLANT _____

WEEK No.	HRS/ WEEK	PRODUCTION MAIN PRODUCT Tins/week (15kg)	BY-PRODUCT Tins/week (15kg)	SALE MAIN PRODUCT Bags/week ( )	BY-PRODUCT Bags/week ( )	REVENUE CASH RECEIVED Kwacha/ week	INPUTS RAW MATERIAL Bags/week ( )	FUEL Litres/ week	LABOUR Total working hours/week	MAINTENANCE /REMARKS
Continue	421	4,631				175,978		1,389	505	
10	36	396				15,048		119	43	
Sub-tot	457	5,027				191,026		1,508	548	
11	48	528				20,064		158	58	
Sub-tot	505	5,555				211,090		1,666	606	
12	38	418				15,884		125	46	
Sub-tot	543	5,973				226,974		1,791	652	
13	45	495				18,810		149	54	
Sub-tot	588	6,468				245,784		1,940	706	
14	41	451				17,138		135	49	300hrs Maintenance **Next 929 hrs**
Sub-tot	629	6,919				262,922		2,075	755	
15	38	418				15,884		125	46	
Sub-tot	667	7,337				278,806		2,200	801	
16	35	385				14,630		116	42	Changing of V-belts
Sub-tot	702	7,722				293,436		2,316	843	
17	47	517				19,646		155	56	Turning of beaters
Sub-tot	749	8,239				313,082		2,471	899	
18	41	451				17,138		135	49	
Sub-tot	790	8,690				330,220		2,606	948	

INSTALLATION & SAFETY OF MILLING

8

Installation and Layout of Equipment

Basic requirements

Functionality, strength, safety, security and reliability are paramount when installing mills. Mills must be designed and constructed to be strong enough to operate safely and efficiently at the required capacity. Wherever mills are installed, it is essential to provide enough room that permits space for cleaning and maintenance, feeding of grain and collection of milled products. There must also be enough space and shelter available for cleaning grain and protecting the customers from the weather. In commercial milling centres, storage facilities must be provided for both grain and milled products. This must be under cover, lockable, dry and secure from all pests. Figure 37 illustrates some floor plans for different milling options drive by diesel engines.

Hand-operated mills

Human-powered milling machines are most commonly housed within a wooden or steel framework. They are relatively lightweight and easily transportable for use wherever required. For comfort and greater working efficiency, the machine should be under cover and sheltered from the direct heat of the sun. Common problems with operating steel-plate mills are the fixing arrangements which can easily break down and render the mill useless. Ensure that a heavy table or structure is used and that the mill is firmly bolted to it and well supported during heavy milling. The damage occurs when free play develops.

Animal-powered mills

A small movable thatched roof structure is enough to cover the milling parts. However, animal-powered machines will also benefit from shade, for example, by installation under trees. If the mill is shared, it would be best to build it in a central and commonly owned place to avoid any disagreements. It is also important that users agree on the utilization of the animal(s). This varies greatly from village to village. In some places, families use their own animals while in others, a women's group or village association will provide use of an animal collective. In that case, it is important to determine how responsibility of the animals should be shared.

Motor-powered Mills

Diesel engine

Diesel-powered mills must be firmly bolted on to a fixed mass such as a thick concrete plinth (15–20cm) or floor to minimize problems caused by excessive vibration. The ground underneath the plinth must be properly rammed and preferably filled with stones before casting the concrete. The foundation must be broader than the frame of the engine/mill (30cm on each side). Ideally, the concrete foundation should be located within a building where there is protection from the weather and to prevent damage being caused to the motors. Except for the plinth, many rural mills have just earth floors. This is, however, not recommended due to hygienic reasons.

The building also provides a more organized environment in which to carry out maintenance and repair work. The building can be wholly or in part open-sided as good ventilation and light is important in a mill. Walls can be made out of many materials. Timber stuck into the ground is one type of simple walls while clay bricks, burnt bricks or cement blocks are more common. If a combustion engine is used, it is recommended to have a fireproof roof such as corrugated iron or asbestos cement sheets.

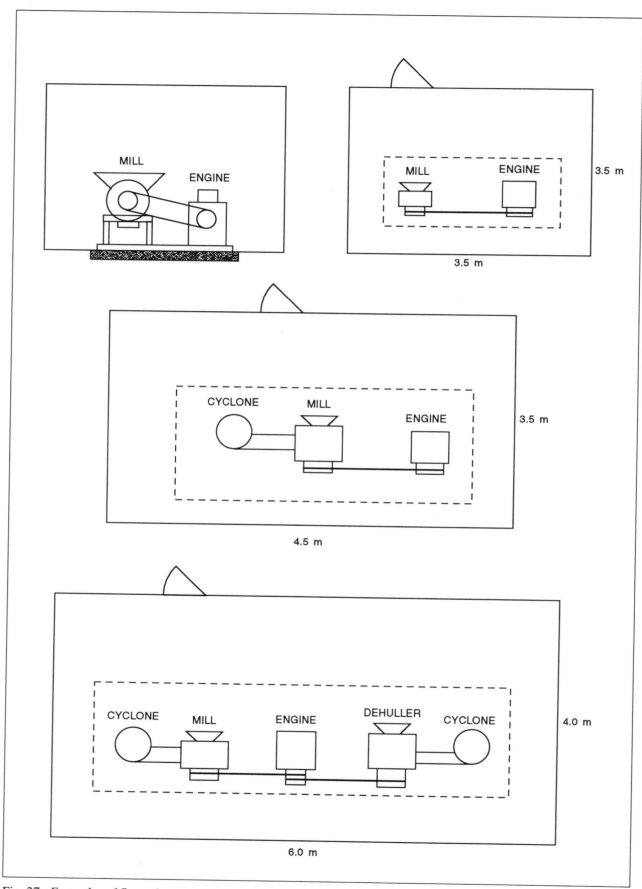

Fig. 37 *Examples of floor plans*

Electric motor

When an electric motor is to be used, one must always make sure that the correct type (voltage, phases) of electricity supply is available and that all power cables and fuse boxes are fully insulated and protected from rain. A thatched roof of good quality can be used for a mill driven by an electric motor. Due to less vibration, the foundation for en electric-driven mill does not require the same thickness and strength as with a diesel engine. A licensed electrician must approve the installation.

Steam engine

Steam-engine-driven mills are hardly an option for rural small-scale milling and require specialist competence for design and construction.

Water-powered Mills

Water- and wind-powered machines must be sited at the most convenient position for harnessing the source of power. This location may be a considerable distance from the village. The resulting inconvenience and wasted travelling time would detract from the usefulness of such a system. For this reason communities sometimes congregate around a water-mill.

Due to the requirement for constant flow, the viability of water-mills is highly dependent on the topographic formation of the area (site-specific) and demand expertise in design and construction. If a suitable location is available, it is important to ensure that the installation does not come in conflict with irrigation and other needs. Mill-ponds cannot be constructed easily on hard rock or an permeable soils. Mill-races, however, can be cut through most rock and soil, and can be lined with concrete or bitumenized wood.

Wind-powered Mills

Wind is an unpredictable source of power and a storm can force a large amount of additional load upon a windmill and its supporting structures. Being exposed to the weather, careful thought at the design stage is necessary to produce a windmill capable of withstanding all such events.

Before installation of a windmill, a precise evaluation should be made of wind potential, including variability. Wind conditions are highly site-dependent. Sheltered areas (er.g. valleys, near to forests, buildings) are generally unsuitable, whilst hilltops are more ideal. It is best to install the windmill on top of a tower several metres high. Wind speed at this height is both greater and more even because it is not affected by any irregularity (trees, buildings) found at ground level. However, this increases the likelihood of mechanical vibration problems and storm damage, and hence highlights the need for careful design and construction of the mill.

Safety

Rotating machinery is potentially dangerous and hence safety is a most important consideration when working on any grain mill. All grinding wheels, belts, pulleys, chains, shafts, etc., should be guarded to prevent limbs (including fingers) and pieces of clothing from becoming trapped and causing an accident. Figure 38 shows the danger of unprotected transmissions. Many people have been killed or crippled unnecessarily due to safety negligence. The cost of making extra protection on a mill is marginal. Chapter 7 discusses the basic safety precautions for a diesel-driven hammer.

Great care must be taken to prevent foreign debris (sticks, stones, mud, metal, etc.) from entering the mill as these can damage the grinding wheels and their drive gear. It is possible that a grinding wheel could explode if a large stone were accidentally fed in with the grain. This type of accident would not only be expensive but could also cause serious injury. To prevent such an incident, the wheels should be enclosed behind strong guards.

Before starting any machine all moving parts should be examined to see that they are both secure and unlikely to come loose or fall off during operation. Regular maintenance and tightening of bolts and nuts are important parts of safety management. Always be sure to switch off the power and wait for the mill to stop rotating before attempting to carry out any maintenance. Always disconnect the power when the mill is not in use.

In almost any country there are strict laws against removing protective guards. It is a criminal offence to take them away during milling and it is the

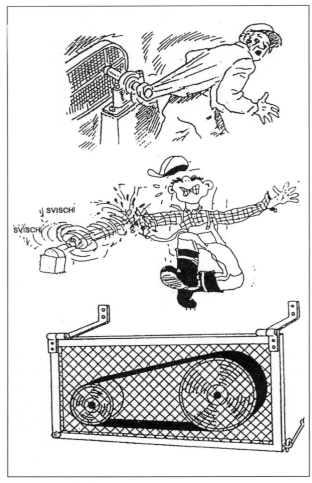

Fig. 38 *Danger of unprotected transmission*

policy should be to avoid dust formation as much as possible, for instance by attaching a cyclone to the mill which reduces dust problems substantially.

Fig. 39 *Dust*

Keep children away from the mill!

The risk of fire is a serious problem in most mills. The basic precautions are:

- never store fuel in a mill; store it in a safe place away from the mill;
- use a fire-resistant roof;
- repair leaks on fuel lines; use a funnel when filling fuel and oil;
- no smoking or open fire in the mill;
- never use water to extinguish burning fuel; suffocate it by covering with a blanket;
- cleaning!
- keep materials which easily catch fire away from the mill.

responsibility of the operator to ensure that protections are firmly in place.

If during operation the local air becomes full of dust/flour it is advisable to wear a mask to filter these particles (see Figure 39). A cloth covering the mouth and the nose is, however, not a satisfactory protection against inhaling dust. Old and mouldy dust can be very dangerous to the health. It may also prove more comfortable to wear eye goggles when using the mill. High concentrations of flour dust are explosive; thus a production area should be free from all potential sources of ignition. A general

Another neglected risk in many mills is contemination of flour. Fuel on the floor can be absorbed by the flour. Old and mouldy dust can cause poor taste. Diluting good grain with cheap but mouldy grain can be tempting in small commercial mills but it is a highly dangerous practice due to the toxicity of many moulds. Never allow surplus and treated seed to enter the mill to be used for human or animal consumption. Ensure that the customers know how to use storage chemicals in a proper way.

SUMMARY OF MILLING PROBLEMS

9

If a mill is kept in good condition it will operate efficiently and have a longer working life. A mill therefore needs regular inspection and maintenance to keep it in good working order. Case Study 3 gives a detailed description of maintenance issues for a diesel-driven hammermill. The sources of power varies and their maintenance requirements are make and model specific. Therefore one has to study their specific fault finding and maintenance recommendations. The intention of this chapter is to concentrate on the milling aspects and not the source of power. A general milling operation trouble chart is presented in Table 7 on the next page.

Below are some general maintenance tasks that will be required by most types of milling machines to ensure minimum problems.

1. Check the strength and security of the machine structure ensuring that all moving parts are correctly and safely assembled.

2. Check that all machine guards are in place.

3. Lubricate according to instructions any gears, chains, bearings, brushes, etc.

4. Clean the grinding wheels and mill interior regularly.

5. Check that the belt/chain drive tension and alignment is correct.

6. Read the handbook (instruction manual) attached to the mill and follow the recommended operation and maintenance recommendations provided.

If a combustion engine is being used as a source of power it is important to have a skilled mechanic available in the vicinity to assist the operator beyond basic maintenance to keep the engine in good working order. The same can be said about a good electrician to ensure that the electric installation is in good shape. Water- and wind-powered mills are designed and constructed according to local specifications and require locally trained operators and mechanics to maintain and operate these machines.

Table 7 is a guide to possible milling troubles which may be encountered during the operation of a mill.

Table 7. MILLING OPERATION TROUBLE CHART

DEFECT	MAY BE CAUSED BY	REMEDY
Will not turn	No power supply	Check supply - ensure motor or other source of power is working correctly
	Too much fixed load	Disconnect power, check to see if over-feeding has caused a jam, check screen for blockage
Mill will not turn fast enough	Not enough power	Check if power source is working correctly
	Too many machines connected	Remove belt drive from additional machines connected to power supply
	Over-feeding	Reduce grain feed rate
Noisy operation	Worn bearings	Check bearings and replace or lubricate as required
	Loose machine parts	Check tightness and security of the assembly
	Rubbing action between machine parts	Check alignment of shafts and other moving parts
		Check clearance gap between grinding wheels and/or shear plates/knives
Mill does not rotate freely while empty	Drive shaft or bearing problems	Check bearings and shafts for distortion and/or wear
Inadequate dehusking/dehulling (dehullers only)	Inadequate rubbing/impact action	Repair, reset or replace worn parts (drums, discs, rubber rollers, friction area, shear plates, etc.
	Crushed grain	Reset clearence between rubbing /cutting parts
		Ensure right moisture content of grain for the machine
	Excessive rubbing action	Reset and clean blocked screens or outlets
Low output	Caking/blockageing in the mill	Clean and scrape off inlet, milling chamber, between grinding wheels, screen and outlet
		Dry grain if too wet
	Stones/wheels polished	Remove for resurfacing or replacing
	Inadequate crushing action	Turn and/or replace beaters
	Inadequate crushing action	Moisten grain if too dry
	Worn parts	Replace or repair
Coarse grinding	Stones/wheels polished	Remove for resurfacing
	Inappropriate feeding	Adjust feeding device (smaller quantity)
	Damaged screen	Repair and/or replace
	Worn beaters	Turn and/or replace beaters
Hot flour	Over-grinding of material	Check clearance gap between grinding wheels
		Partly blocked screen
	Over-feeding	Reduce feeding with over-sized source of power
Belt slips - remains on the pulley (flat belt)	Under-rated power take off	Connect a larger belt (width) or a second belt
	Over-rated power take off	Too many machines connected to power supply – disconnect drive to one machine
	Stretched belt	Check belt tension and adjust machine position or shorten belt as necessary
	Too much fixed load	See above
	Polished surface of pulleys	Roughen surface for grip or use dressing paste (belt-lap, tar)
Belt slips (V-belt)	Under-rated power take off	Change to pulley with more belts and/or increase number of belts
	Stretched belts	Check belt tension and adjust machine position
		Uneven length of belts – replace whole set of belts
Belt slips from the pulley	Inaccurate alignment of machine	Adjust position of machine, do not use a stick or similar to keep the belt in position.

SMALL-SCALE MILLING MACHINERY

SMALL-SCALE MILLING MACHINERY   **10**

The following tables (8–13) are designed to give a general view of small-scale cereal milling equipment. They are by no means exhaustive and are made possible by the response of the manufacturers listed to requests for information. Details of the machine capabilities are for general guide purposes only and will be affected by condition of mill, operational system and grain quality.

The outputs given by manufacturers are from well-managed mills and it is realistic to use them as maximum production figures which can be used for selecting the best alternative but also for assessing the performance of mills.

Table 8. HUSK REMOVAL

MANUFAC-TURER	TYPE	OUTPUT kg/hr	HORSE-POWER	REMARKS
Alvan Blanch	Steel huller manual	15		
	Steel huller	250-350	12-15	
	Rubber roll	1,000	5.5	
CeCoCo Ltd	Centrifugal manual	250		
	Rubber roll	360-1,200	0.5-5	
Colombini	Steel huller manual	14		
	Steel huller	160-350	12-15	
Dandekar machine works	Rubber roll	900-1,100	3-5	
Kisan Krishi Yanta Udyog	Centrifugal	400-500	2	
	Rubber roll	1,500-2,000	10	
Lewis C Grant Ltd	Steel huller	140-300	12-15	
McKinnon	Steel huller	40-310	5.5-16	

The power requirement is quoted in horsepower in line with manufacturers' specifications.

To convert: Multiply by
Horsepower to Kilowatts   0.746
Kilowatts to Horsepower   1.34
Remarks:
     D = diesel   E = electric   T = tractor (PTO)
     H = hand

Table 9. BRAN REMOVAL

MANUFAC-TURER	TYPE	OUTPUT kg/hr	HORSE-POWER	REMARKS
CeCoCo Ltd	Mobile	10-15		
	Abrasive cylinder	300-12,000	5-30	
Dandekar Machine works	Abrasive cone	1,200-1,700	6-15	
	Friction jet	1,200-1,700	15	
Kisan Krishi yantra Udyog	Abrasive cone	400-16,000	5-20	
	Abrasive cylinder	400-700	7.5-15	
	Abrasive disc	200-300	3	
Lewis C Grant Ltd	Steel huller	270-590	12-15	
McKinnon	Steel huller	270-590	9-14	
PRL/RIIC		250-500	8	

87

Table 10. HAMMERMILLS

MANUFAC-TURER	TYPE	OUTPUT kg/hr	HORSE-POWER	REMARKS
Alvan Blanch		110-1,250	3-25	E/D/T
Ateliers Albert		1,500	20	E/D
Baldeschi & Sandreani		150-1,700	10-30	E/D
C S Bell		680	2-19	E/D
Christy Hunt		64-1,500	3-20	E/D
Cormall		85-505	7.5-12	E/D
Electra		80-800	2-14	E/D/T
Gosling Group		35-1,800	5-40	E/D/T
Ind. Maquina		80-1,500		
Law Denis		270-1,250	7.5-15	E
Manik Engin.		90-816	8-30	E/D/T
Mio Osijek		50-200	0.7	E
Ndume Ltd		190-1,270	8-25	E/D/T
Penagos Hermanos		90-1,500	3-8	E/D
Philco Dierings		317-907	15-20	E/D
President		90-1,500	15-20	E/D
Skiold		80-1,350	7.5-80	E/D/T
Tradepoint	Complete container mill	800-1,000		

Table 12. PLATE MILLS

MANUFAC-TURER	TYPE	OUTPUT kg/hr	HORSE-POWER	REMARKS
ABC Hansen		6-18	0.5-1	E/H
Alvan Blanch		15-400	0.5-10	E/D/H
Christy Hunt		7-275	1-8	E/D/H
Cormal		250-350	5.5	E/D/H
Ndume Ltd		15		H
Penagas Hermanos		200-400	3-6	E/D
President		250-1,100	3-15	E/D/H
Skiold		150-350	5.5-11	E/D/H

Table 13. ROLLER MILLS

MANUFAC-TURER	TYPE	OUTPUT kg/hr	HORSE-POWER	REMARKS
Baldeschi & Sandreani		400-600	12-15	E
T W Barfoot	Container mill	1,000 - 2,000	60	E
Favini Impianti	Mobile/ container mill	600- 1,000	120	E
Maize Master		up to 500	10	E
Ocrim SpA	Mobile/ container mill	500-583	70-120	E/D
President		400-1,400	3-10	E

Table 11. STONE MILLS

MANUFAC-TURER	TYPE	OUTPUT kg/hr	HORSE-POWER	REMARKS
ABC Hansen		100-1,000	3-25	E
Alvan Blabch		150-1,000	7.5-18	E/D
Dandekar Machine works		200-300	7-10	D
Ets Guy Moulis Constructeur		50-800	1-6	E/D
Skiold		250-1,000	5.5-15	E/D

MANAGEMENT 11

Management is a broad subject and this chapter will deal mainly with the administrative and financial matters of establishing and operating grain mills.

Introduction

Management is a vital tool in planning and developing sustainable milling operations. A management system must be designed to cope with the local situation. Inflation rates above 100 per cent per year and interest rates of 30–50 per cent are common in many African countries. Coping with these factors demands a management system which is sensitive to price changes in order to maintain profitability in real terms. After a general discussion on management issues there will be a case study on a management system for small-scale milling promoted in Zambia by the co-operative movement. They have promoted rural small-scale milling (credit and extension) for a number of years and gradually improved the management system with the aim of maintaining sustainable milling operations. The system promoted will of course be highly relevant for most developing countries.

Mechanized milling will require technical services and training of millers and operators in management. The best people to perform this function ought to be the Technical Extension Agents (TEAs) working with improved rural productivity. The extension responsibilities in the field of milling are broader than what is normally involved in a technical project. It covers:

- technical aspects,
- economical aspects,
- social aspects,
- nutritional aspects.

It is essential that TEAs cover these fields in their extension work particularly as mechanized milling is a sensitive gender issue.

One can distinguish between three levels of management:

Level 1 Operator's level (in charge of operation, record keeping);

Level 2 Owner's level (Level 1 plus accounting, basic costing, management);

Level 3 Advisors/TEAs level (Level 1 and 2 plus planning, marketing analysis, profitability projections, cash-flow analysis, management systems, etc.).

Ideally, the owner should have the same level of experience as the TEA. However, this is a long-term target as many mill owners lack formal schooling and experience to cover Level 3 within the foreseeable future. During the initial stage it is the responsibility of the TEA to assist and train the entrepreneurs in the listed tasks. The case study will deal with practical examples at all three levels and it will cover:

- Daily Records Form (operator, owner);
- Weekly Records Form (operator, owner);
- Cash Book (cashier, owner);
- Processing Project Description (owner, TEA);
- Profitability Projection (owner, TEA);
- Cash-flow Analysis (owner, TEA).

Sometimes there is a cashier attached to a mill and his/her main responsibility is the collection of milling charges and money from selling flour, payment of inputs, recording and basic accounting. In his/her case, the TEA will have to choose selected training topics from Levels 1 and 2.

How to choose a mill

Before considering the establishment of a small-scale mill, it is essential that a thorough assessment of its needs are made. This involves examining the present system of cereal processing, the demand and the effect of any alternatives, improved method and the likely demand for the mill's services. It is necessary to find out who is going to use the mill before selecting the type and scale of mill suited to their needs.

A small-scale survey should be carried out to examine market demand. It is useful to design a logical questionnaire to cover the necessary issues. A check-list of questions is given below to assist in checking the socio-economic feasibility of a mill. The project description should cover the following points:

1.   How many people are living in the area within specific distances from the planned mill (5km, 7.5km and 10km)? Total quantity milled in the area? Interest in using custom milling services (dehulling and grinding)? Interest and ability to pay milling charges? Quantity of grain to be milled at customs mill? Potential for commercial milling?

2.   Distance to mill of competitors? Flour quality and milling charges of competitors? Rate of competition of customers? Effect of competition?

3.   How big should the mill be? Is it likely that the capacity will be fully utilized given local cereal production and milling demand? The use is likely to be seasonal.

4.   For the size of mill decided above, what is the cost of all the milling machinery and associated equipment needed, and the cost of constructing a building to house it? The latter must include foundation, stores if required, winnowing space and paths, and power lines and transformers if electric motors are to be used. The cost of construction of wind and water power must be added to the mill costs. Will any of the investment be communal?

5.   What will be the labour, fuel and input demands of the mill and can they be met? Will it be possible to maintain a regular supply of fuel or electricity, i.e. are there periodic fuel shortages or electricity cuts? How much cash (working capital) will be needed to operate the mill throughout the year?

6.   What is the traditional method of milling used in the area? Will the mill fit into the local scene: would a new system be preferred – by those who at present mill the grain and those who consume the products? Will these be as acceptable and nutritious as traditional products (ability to separate different nutritious parts of the grain and fermentation)?

7.   Reduction in labour demand for household milling tasks is a likely consequence of a new milling system. Will this free women from other farm and household tasks and offer them new opportunities? Effect of milling charges. How many can afford customs milling? Will the new mill create employment opportunities?

8.   Is there sufficient mechanical and commercial skill to operate a mill? Can the milling equipment be maintained locally and are spare parts available? Is a good supply of spare parts and competent servicing skills available? If spare parts have to be ordered from abroad, the time delay and possible difficulties in obtaining foreign exchange may force the mill to close.

9.   Will credit be available to finance the mill? At what interest rate? Repayment conditions? Are these affordable?

10.   What returns are likely from realistic estimated capacity use? What is the estimated payback period or time to cover the cost of the loan? Are there systems developed for adjusting milling charges to match inflation?

11.   Who benefits from the mill? Nutritious considerations? How will these benefits spread? Who will own the new mill? How capable are the owners and what relevant experience do they have? How committed are they to the success of the mill and how much of their own resources are they prepared to put in to the venture? Will the owner or owners receive a fair return, given the risk and returns from alternative investments? Will men take over the control of milling from women?

12.   Is the traditional marketing mechanism going to be changed by the new system?

13.   How are the by-products utilized?

Comments

Accessibility is the major criterion for selection of an appropriate mill location. There must be access for inputs from producers and outputs to consumers – in the case of customs mills, these are the same people. Commercial mills generally have been more successful in areas accessible to urban markets than in remote rural areas. Being close to the demand is more important than being close to areas of production because product quality is retained better as grain than as flour or meal, when rancidity can develop in a short time. Also grains are easier to transport than flour. An inadequate supply of grain due to limited access – poor transport, lack of a market – will prevent mills from operating successfully. Transport must also be easily available to maintain fuel supply. If the inputs come from several villages, year-round access may be required unless large stores are established. The distance from competing mills must be taken into account in determining location and market structure.

The choice of the scale of production is limited by investment funds available and by the local need for milling. It is also limited by the size of machine available as small mills are manufactured in a limited range of sizes, as are diesel and electric motors.

For mills of a capacity up to 1.5 tonnes per hour the potential market area may be a group of farms, villages or an urban area. For smaller mills it may be the village in which it is located alone. It is necessary to estimate the size of demand for meal or flour in the area and this requires some consumption information. This may be acquired either from secondary sources (consumption surveys and population data) or from a market study. The entire production of an area will not be processed at the proposed mill because some people will still continue to use traditional methods while others will prefer alternative mills. Usually a surplus will be marketed as grain. It is necessary to estimate local consumption and the number of farmers who are likely to pay for the service of a customs mill.

Once potential demand and market share have been estimated, it is necessary to investigate the supply of grain. Over-estimation of demand for the mill's services will cause under-utilization and possible failure. Total volume and seasonal fluctuations should be estimated as they can vary widely through the year according to storage capacity. A customs mill may process a large amount in a relatively short time after the harvest period and be less busy for the rest of the year; if there are two har-

vests a year, there will be two peaks in milling activity. Storage capacity at the mill or on farm will alleviate the peak in demand for milling. Most farmers will store their cereal as grain on the farm and take their requirements to be milled only when the need arises, thus spreading demand through the year.

Financial Costing

Once all costs and benefits have been established one must assess the feasibility of the project.

Two simple methods can then be used, in combination, to assess feasibility; the pay-back method and budgeting procedure.

Pay-back method

The pay-back method simply determines how long it takes before the original capital investment is 'paid back' or the cumulative net returns exceed the investment. It is the number of years before the initial cash investment is expected to be equalled by future cash inflows.

The pay-back method is an indicator of risk (the faster the pay-back the less the risk) and will ensure that returns cover investment costs. The limitation of using this method is that it does not take account of cash flows expected after pay-back such as the need to reinvestment capital to maintain operation or measure the overall profitability of the investment.

Budgeting procedure

The other approach of financial costing of a mill is to use a simple budgeting procedure. In addition to financial costing, this system can also be used for adjusting milling charges under rapidly changing conditions in order to keep business intact and to follow up the actual result. The different components of the budgeting procedure are described below and they will also be exemplified in Case Study 4: Management, which is based on actual figures from Zambia.

1.   Calculate annual output, in kilos, tins or

tonnes of cereals processed. The capacity of the mill will determine the maximum possible output, but in practice output will be less due to the seasonal production fluctuations, than the irregularity of grain supplies and stoppages for repairs and maintenance.

2.   Estimate capital costs, including the cost of milling and other equipment, building costs and land, and if necessary, a cash fund or 'working capital' for operating the business. These are major items of expenditure for a milling operation. Such accurate estimates or quotations are needed for all the various pieces of equipment.

3.   Normally one estimates capital costs on an annual basis, that is to say the cost corresponding to each year of operation should be calculated. This can be done most accurately by using annuity tables, but for those unfamiliar with them, annual capital costs should be calculated as the sum of the interest cost and depreciation. With a high inflation rate, costing on annual basis can be misleading and it is necessary to make quarterly adjustments. The interest cost should be calculated irrespective of how the mill is financed (i.e. whether the mill is financed by a loan or by the miller's own funds), and should be set at a level which satisfies the lender's and the owner's minimum requirements for return on capital.

Depreciation is also an important cost, and must be calculated for equipment, vehicles and buildings which lose value during the life of the project. The simplest method of depreciation is 'the straight-line method' by which: (a) the value lost during the project life is calculated as the difference between the initial value and the residual value, if any, at the end of the project, and (b) the value lost in a year which is calculated by dividing the value lost by the years of 'useful life' expected of the equipment or the building concerned. Equipment is often assumed to have a useful life of 10 years, vehicles 5–8 years and buildings 25 years with good management, but different assumptions should be used depending on local circumstances. For example the average lifespan of a diesel-driven hammermill in Zambia is only 4–5 years. It is essential to base calculations on actual conditions and not 'dreams'. Land does not normally fall in value, but often appreciates, and therefore has no depreciation cost.

4.   Annual operating costs should be calculat-

ed. These include raw materials, labour, power, maintenance and spare parts, stationery, etc. Quarterly adjustment is required on price levels when the inflation rate is high.

Annual fuel costs are estimated by multiplying the fuel price, hourly consumption and estimated number of hours per year. To calculate the annual cost of electricity the product of the unit charge per HP (kW), the rated HP (kW) of the installed motors and the estimated number of hours of operation per year will be added to the annual standing charges.

Labour costs, like energy costs, depend on the hours and days of operation as well as capacity and number of milling operations. Labour will be required for moving and handling grain as well as operating the mill. Costs will vary according to the need for skilled or unskilled workers. In a small customs mill labour requirements beside that of the operator (or owner) may not be large.

5.   Total annual costs should be calculated and compared to the expected revenue from milling fees and the sale of products and by-products, to assess profitability. In the way it is calculated in Case Study 4, profitability refers to profits over and above the minimum return required for the miller to enter the business.

The same calculations should also be carried out on a unit basis, as this helps in fixing milling charges, and allows one to compare the performance of different mills. Assuming that there are no differences in convenience and product quality, the milling system with the lowest unit production costs is likely to be most appropriate. With customs mills, the mill with the lowest depreciation, interest and energy costs will probably be the most suitable, as labour and other costs are unlikely to vary widely.

6.   'Sensitivity analysis' should be carried out to determine how costs and profitability are affected by varying critical assumptions, for example by reducing annual output by 25 per cent and by increasing capital costs by 15 per cent. This helps in assessing the risks facing the project, if it does not perform as originally assumed.

Follow-up

The system of making profitability projections can also be used to follow up the performance of a mill by adjusting the actual figures quarterly or at end of the year. By combining the planning system with the actual result system one gets an excellent tool for the owners and the TEAs to assess the correlation between planning and actual result and a tool for monitoring and evaluation of performance. The described system is especially useful for adjusting the milling charges.

The system of taxation varies for each country but the proposed follow up systems of small mills can be used as under most conditions.

CASE STUDY 4

Management

Introduction

This case study on management is based particularly on the experience of the co-operative movement in Zambia, but it can be applied to most milling conditions. The figures are taken from a real case from the beginning of 1992. The prices change rapidly, however, due to inflation and should be considered as an example only.

In Case Study 3 – Operator's Manual for Diesel Powered Hammermill – there is a section discussing management aspects at operator's level which is also connected with regular maintenance of the mill and its engine. The sections on daily and weekly recording will be repeated in Case Study 4 – Management – in order to cover all the different management levels.

The system of cost calculation is a good tool in selecting the most appropriate mixture of milling unit and source of power, i.e. what is most appropriate: a simple mill with a big engine or a more sophisticated mill with a smaller engine? With the help of test reports from the Magoye Agricultural Engineering Centre, it is possible for the TEA to assist the milling units in the selection of the best and most economical option. Another aspect is that the TEA can also influence manufacturers and dealers in providing the optimum choice of machinery.

The management forms used cover both grinding and dehulling. They also include both customs and small-scale commercial milling units. The user will just have to select the right wording in the headings in order to cover the right products and activities. Customs milling means that the households grind their own grain in the mill and pay only the milling charges. A commercial mill buys the grain and sells flour and bran. In many cases there is a combination of service/customer processing units. In fact this is a recommended practice as it will increase the use of machinery and generate more income.

Functioning management systems are essential at all levels as they affect each other. It is the work of the TEAs with the assistance of the Farm Management, Home Economics and Nutritionists Staff to ensure that necessary training and extension services reach all levels in improved grinding and dehulling.

The ideal situation is when mill owners are capable of covering all management aspects in an efficient manner. However, only few have the capacity today and they need assistance with particularly cost calculations and cash-flow projections. This is a field where also many TEAs have weak experience and it is essential that management aspects play a major role in the in-service training programme on processing at village level. In practical field work the TEA will have to adjust according to their customers' knowledge and most owners are unfortunately not capable without training to go beyond the operator's level.

Experience shows that it is more appropriate at field level to use record-keeping books for data instead of loose blank forms which get damaged easily. However, with calculations in an office it goes faster with pre-printed blank forms. The sets of cases are examples of how the different forms can be designed.

CASE 4.1	**DAILY RECORDS FORM**							
YEAR: 1992	MONTH: February		TYPE OF PROD.: Customs milling				OPERATOR: J.K. Banda	
WEEK (DATE): 17/2-23/2	WEEK NO.: 8		NAME OF PLANT: Kazimuli					
PRODUCTION			SALE		REVENUE	INPUTS		
DAY HRS/ DAY	MAIN PRODUCT Tin (15 kg)	BY-PRODUCT Tin (7.5kg)	MAIN PRODUCT Bags ( )	BY-PRODUCT Bags ( )	CASH RECEIVED Kwacha	RAW MATERIAL Bags ( )	FUEL Litres	LABOUR Initials and hrs worked/day	OTHER INPUTS, MAINTENANCE, REMARKS
MON 6 HRS	60				2,280		18	BB 7 CC 3	
TUE 5 HRS	50				1,900		15	BB 6	
WED 7 HRS	70				2,660		21	BB 8 CC 3	220 l diesel drum
THU 4 HRS	40				1,520		12	BB 5	
FRI 7 HRS	70				2,660		18	BB 8 CC 2	1/2 oil added Tighten fan belt
SAT 7 HRS	70				2,660		18	BB 8 CC 3	
SUN 0 HRS	0				0		0		
TOTAL 36 HRS/WEEK	360				13,680		102	BB 42 CC 11	

Daily Records Form

It is recommended to use record-keeping books where appropriate headings are listed (see example on next page). The data must be filled in daily if the operation system is to have a production/sales data collection. This information will be required for accounting and cost follow ups. The owner must ensure that the operator or somebody attached to the mill fills in this form in a proper way.

The data listed in the form are examples of what to include. This example covers customs milling.

More information is required for commercial milling like sales figures and inputs. The main product can be flour and the by-product bran.

Field visits show that many mills have poor and incomplete records. By neglecting record keeping, the ability for proper management is seriously reduced. A processing unit not capable of keeping basic records is hardly a sustainable enterprise.The Daily Records Form will make accounting easier and more accurate. It will also shows how well inputs have been used and indicate mismanagement.

WEEKLY RECORDS FORM

CASE 4.2

OPERATOR'S FORM

YEAR: 1992 | TYPE OF PRODUCTION: Customs milling | OPERATOR: J.K. Banda

WEEK No.	HRS/ WEEK	PRODUCTION MAIN PRODUCT Tins/week (15 kg)	PRODUCTION BY-PRODUCT Tins/week (15 kg)	SALE MAIN PRODUCT Bags/week ( )	SALE BY-PRODUCT Bags/week ( )	REVENUE CASH RECEIVED Kwacha/ week	INPUTS RAW MATERIAL Bags/week ( )	INPUTS FUEL Litres/ week	INPUTS LABOUR Total working hours/week	MAINTENANCE, REMARKS
Continue	421	4,631				175,978		1,389.3	505.2	
10	36	396				15,048		118.8	43.2	
Sub-tot	457	5,027				191,026		1,508.1	548.4	
11	48	528				20,064		158.4	57.6	
Sub-tot	505	5,555				211,090		1,666.5	606	
12	38	418				15,884		125.4	45.6	
Sub-tot	543	5,973				226,974		1,791.9	651.6	
13	45	495				18,810		148.5	54	
Sub-tot	588	6,468				245,784		1,940.4	705.6	
14	41	451				17,138		135.3	49.2	300hrs Mainten. Next 629hrs
Sub-tot	629	6,919				262,922		2,075.7	754.8	
15	38	418				15,884		125.4	45.6	
Sub-tot	667	7,337				278,806		2,201.1	800.4	
16	35	385				14,630		115.5	42	Changing of V-belts
Sub-tot	702	7,722				293,436		2,316.6	842.4	
17	47	517				19,646		155.1	56.4	Turning of hammers
Sub-tot	749	8,239				313,082		2,471.7	898.8	
18	41	451				17,138		135.3	49.2	
Sub-tot	790	8,690				330,220		2,607	948	

Weekly Records Form

The Weekly Records Form on the following page adds up what happened during the week. The operator includes the weekly summary which will be added to the previous week. There should be an accumulated sub-total every week. This is crucial for knowing when there is time for service and to determine efficiency of operation.

The information gathered in the Weekly Records Form is a great asset both for the owner/manager and the TEA as it provides an instant indication of production result. By regular checking, the TEA will have an accurate tool for determining costs and profitability. It will also indicate weaknesses/ strength in the operation.

CASE 4.3		CASH BOOK							
CASH BOOK NO.: 2			YEAR: 1992		PRODUCTION UNIT: Kazimuli				
No.	DATE	ACTIVITY	CASH IN	CASH OUT	BALANCE	BANK IN	BANK OUT	BALANCE	PRIVATE
		Balance brought forward			33,302			133,303	25,000
33	3/3	Milling 1/3-3/3	4,180		37,482				
34	5/3	Purch. spares		17,320	20,162				
35	8/3	Milling 4/3-7/3	6,240		26,402				
36	15/3	Milling 8/3-15/3	18,240		44,642				
37	18/3	Purch. diesel (220 l)		2,640	42,002				
38	18/3	Transport		900	41,102				
39	19/3	Deposit in bank		30,000	11,102	30,000		163,303	
40	20/3	Payment of salary BB		2,350	8,752				
41	20/3	Payment of salary CC		1,200	7,552				
42	22/3	Milling 16/3-21/3			21,232	18,240		181,543	
43	25/3	Paym. instalment & interest	13,680		21,232		51,940	129,603	
44	28/3	Spares	30,000	42,000	9,232		30,000	99,603	
45	28/3	Owner, private		5,000	4,232				5,000
46	31/3	Milling 22/3-31/3	15,960		20,192				
47	31/3	Owner private		15,000	5,192				15,000
48	31/3	Repairs	40,000	37,000	8,192		40,000	59,603	
	31/3	To be brought forward			8,192			59,603	45,000

Cash Book

The basic principle for a cash book for milling is CASH IN, CASH OUT, BANK IN and BANK OUT.

With the selected cash book system one will get a good information of the flow of both cash and the money in the bank.

It essential in all accounting to verify all transactions and to keep receipts. A system must be devised to systematically collect and record all activities involving money. It is highly recommended that the owner(s) of a processing unit ensure that an up-to-date cash book is kept in good order.

An accurate cash book is an excellent tool for management of funds. Cash is often in short supply, which makes operation of a processing unit a delicate issue particlularly when credit is involved. When a loan is due for payment there should be money available to pay it in time. A good payment record is a precondition to get credit on commercial terms. Breakdowns occur regularly and funds must be available for spare parts and repair works. A good accounting system increases the readiness to meet unforeseen expenditures.

The normal way of accounting is a cash book and a seperate file for receipts, Daily and Weekly Record Forms. The headings listed in Case 4.3 should be added in the cash book. Recording should be made with a ball-point pen and not a pencil. Deposit of money in the bank will reduce cash and withdrawal from the bank will increase cash. A useful heading in the cash book is money used from the milling plant for private use as it will sometimes be necessary to make a distinction between operating business and private transactions.

It is normally good business for an owner to ensure that the milling plant has somebody capable of keeping cash books in good order.

CASE 4.4

MILLING PROJECT DESCRIPTION

PROJECT: Hammermill	DISTRICT: Chadiza
VILLAGE: Kazimuli	PROJ.PREP.DATE: 15/1/92

1. MILLING PROJECT SUMMARY

TITLE: Diesel-driven hammermill			
PROJECT MEMBER(S):	Primary Society		
PROJECT SITE: Kazimulu, Eastern Province			
ACTIVITIES: Milling			
SOURCE OF FUNDING:	Own 1/3 Co-operative Credit Scheme 2/3		
EXPECTED STARTING DATE/CANCELLATION DATE: 1/3/92			
CAPITAL REQUIREMENT:		LIFE-SPAN	KWACHA
PROJECT ACQUISITION COSTS:			
Mill		4 yrs	316,000
Initial spares		4 yrs	11,000
Transport		4 yrs	9,000
Accessories		4 yrs	3,000
Shed & Foundation		16 yrs	64,000
INITIAL WORKING CAPITAL:			10,000
TOTAL INVESTMENT:			413,000
INITIAL FUNDING REQUIRED:		PERIOD	KWACHA
OWN FUNDS INCLUDING INSTALMENTS			133,000
PRINCIPAL LOAN (interest rate: 38%)		3 yrs	280,000
TOTAL FUNDS REQUIRED:			413,000

2. DEMAND FOR MILLED PRODUCTS

TYPE OF GRAIN		SCORING	CUSTOMERS' PREFERENCE	NUTRI-TIONAL VALUE	NUTRI-TIONAL VALUE
MAIZE		1 (LOW) TO 5 (HIGH)	TASTE	ADULTS	CHIL-DREN
MILLING	WHOLE GRAIN	FERMENTED	4	4-5*	4*
		UN-FERMENTED	3	3-4*	2-3*
	DEHULLED GRAIN	FERMENTED	5	5	5
		UN-FERMENTED	4	3	3
	MARKETED	ROLLER MEAL	4	3	3
		BREAKFAST	4-5	3	2-3

Comments: Best choice for Kazimuli customers
1. Fermented dehulled grain 2. Fermented but hand sieved whole grain (*hand sieved after grinding to remove coarse fibre).
Action: 1. Promote dehulling and fermentation before grinding. Start with traditional practices.
2.Diesel-driven hammermilll is a suitable option for grinding.
3. Include mechanized dehulling combined with fermentation

Milling Project Description

Any start of a milling unit ought to be preceded by a proper planning phase based on sound business principles. Too many enterprises have failed due to weak planning and/or inadequate knowledge of the market. Sometimes the plant has excellent products and technical specifications but not enough customers or the reverse. Many of these mistakes could have been avoided with a bit more careful planning.

The aim of Case 4.4 is to provide guide-lines for relevant questioning to obtain important planning information in table forms in a logical order (see next page). The principle is to start from the beginning and answer the questions as accurately as possible. Add other aspects if found necessary for specific areas. Taste and food preferences are sensitive issues and it is always best to begin with the traditional food habits. Be aware that fermentation is considered an outmoded practice by many promoters of 'improved milling

techniques' despite its many obvious advantages.

Marketing analysis based on reliable statistics and tested assumptions is desired and useful in estimating demand. It is easier with a known product like hammermilling but more complicated with less known activities like village mechanized dehulling.

A common practice in determining demand is to identify bottle-necks, unsatisfied demands and methods and products to rectify shortcomings, and to rate people's preferences for different services/products and desire to buy them. Based on the findings after each table, one will make conclusions and projections on potential services and sales to cover the identified demand. This information will form the basis for tentative project designs which should be costed and compared with available resources.

If the initial feasibility study indicates a good potential then the next step includes a proper profitability projection.

TYPE OF GRAIN		SCORING	CUSTOMERS' PREFERENCE	NUTRI-TIONAL VALUE	NUTRI-TIONAL VALUE
		1 (LOW) TO 5 (HIGH)	TASTE	ADULTS	CHIL-DREN
MILLING	WHOLE	FERMENTED			
		UN-FERMENTED			
	DEHULLED	FERMENTED			
		UN-FERMENTED			
	MARKETED				

Comments:

3. TIME SAVED THROUGH MILLING SERVICE

TYPE OF MILLING Daily requirement of grain to be milled: 3 kg Maize	OPERATION Distance to mill 4 km	TIME SPENT/ HOUSEHOLD (HRS/TIN of 15 KG)	TIME SAVED/ HOUSEHOLD	
			HRS/TIN	HRS/DAY
OPTION 1: TRADITIONAL POUNDING (base line)	DEHULLING	6 hrs	-	-
	GRINDING	10 hrs	-	-
	TOTAL:	16 hrs	-	-
OPTION 2: TRAD. POUNDING & MECHANIZED GRINDING	DEHULLING	5 hrs	1hrs	
	GRINDING	4 hrs	6hrs	
	TOTAL:	9hrs	7hrs	1hr24min
OPTION 3: MECH. GRINDING (winnowing, sieving & no fermentation)	DEHULLING	-	-	-
	GRINDING	6hrs	10hrs	
	TOTAL:	6hrs	10hrs	2hrs
OPTION 4: MECH. GRINDING (winnowing, sieving & fermentation)	DEHULLING	-	-	-
	GRINDING	7.5hrs	8.5hrs	
	TOTAL:	7.5rs	8.5hrs	1hr42min
OPTION 5: MECH. DEHULLING & GRINDING (no fermentation)	DEHULLING & GRINDING	5hrs	11hrs	
	TOTAL:	5hrs	11hrs	2hr12min
OPTION 6: MECH. DEHULLING & GRINDING (fermentation	DEHULLING & GRINDING	6.5hrs	9.5hrs	
	TOTAL:	6 .5hrs	9.5hrs	1hr50min

Comments: Fermentation will increase milling by 18 min/day compared to unfermented alternatives. It will also improve taste and nutritional value.

Option 2 best for poor people.

Option 3 will produce a coarse meal with high-fibre content.

Option 4 will produce a tastier & more nutritious meal than 3.

Option 5 is best for sorghum and millet.

Option 6 will produce a tastier & more nutritious meal than 5.

Ranking: 2, 6, 5, 4, 3 (people's preference of taste).

Common practice: 2 & 3.

Problems to find dehullers.

4. MARKETING ANALYSIS

POTENTIAL CUSTOMERS AROUND PROJECT SITE (NUMBERS)	NO. WITHIN 5 KM RADIUS	NO. WITHIN 7.5 KM RADIUS	NO. WITHIN 10 KM RADIUS
PEOPLE	1,800	2,400	3,200
HOUSEHOLDS	300	400	800
HOUSEHOLDS THAT CAN AFFORD/WANT SERVICE OFFERED (milling (M) & dehulling (D))	M 200 D 150	M 270 D 175	M 500 D 350
HOUSEHOLDS THAT CAN AFFORD/WANT PRODUCT OFFERED (flour (F), meal (M) or grits (G))	F 100 M 300 G 75	F 150 M 400 G 100	F 400 G 800 G 250
HOUSEHOLDS THAT CAN AFFORD/WANT BY-PRODUCT OFFERED (bran)	250	300	700

Comments: 1. Enough customers to operate a hammermill.

2. Enough customers who can pay milling charges.

3. Potential market for installing a dehuller.

4. Sorghum is common, used for beer, need of grits and bran.

5. COMPETITORS

COMPETITORS AROUND PROJECT SITE (NUMBERS)	NO. WITHIN 5 KM RADIUS	NO. WITHIN 10 KM RADIUS	NO. WITHIN 15 KM RADIUS
OPERATING WELL		2	4
OPERATING LESS WELL		1	3

Comments: Competitors mills are badly placed, shortage of fuel.

Frequent breakdowns. Kazimuli mill would be attractive at least when new

6. LEVEL OF PRODUCTION

IDENTIFIED CUSTOMERS AROUND PROJECT SITE (PROJECTION)	PRODUCT A MAIZE	PRODUCT B MAIZE* (Potential)	PRODUCT C	PRODUCT D
1. ANNUAL PROCESSING CAPACITY OF PROJECT (TINS, 15 KG)	30,000	30,000		
2. NO. OF HOUSEHOLDS SERVICED (milling (M) & dehulling (D)	M 250	M 250		
3. ANNUAL QTY SERVICED/HOUSEHOLD (TINS, 15 kg)	67	67		
4. TOTAL ANNUAL QTY SERVICED (2x3, TINS)	M 16,800	M 16,800		
5. HOUSEHOLDS PURCHASING PRODUCTS (meal)		150		
6. ANNUAL QTY PURCHASED PER HOUSEHOLD (tins, 15 kg)		20		
7. TOTAL ANNUAL QTY SOLD (5x6, tins)		3,000		
8. HOUSEHOLDS PURCHASING BY-PRODUCTS (bran)	100	100		
9. ANNUAL QTY PURCHASED PER HOUSEHOLD (tins)	15	15		
10. TOTAL ANNUAL QTY SOLD (8x9)	1,500	1,500		
UTILIZATION RATE OF CAPACITY	customs 0.61	commercial 0.71		

*Potential for second stage inclusion of commercial milling.

Comments: Utilization rate is 0.61. This means that there is spare capacity available of 11,700 tins. There is a market for flour/meal.

Potential: 150 households need 20 tins each totalling 3,000 tins. Will improve utilization rate to 0.71.

Conclusion: 1.There is a market for customs milling and dehulling.
2. Start with grinding (hammermill).
3. When customs mill works well, add commercial milling.
4. Future lies in combined customs/commercial milling.
5. Add dehulling.

7. KNOWLEDGE AND SKILL IN MILLING

SKILLS LEVEL GOOD = G AVERAGE = A (NEED ADVICE) POOR= P (NEED TRAINING)	MANAGEMENT	OPERATION	MAINTENANCE	REPAIRS
OWNER(S)	P	A	A	P
OPERATOR(S)	A	A	P	P
ADMIN./ CASHIER	P	A	P	P
MECHANIC(S)	-	G	G	P

Comments: There is a need for TEA to arrange for management training at all levels and training in maintenance and repairs. Inadequate capacity to undertake repairs.

8. AVAILABILITY OF INPUTS

ITEMS	AVAILABILITY/ QUANTITY	NO. OF SUPPLIERS	DISTANCE TO SUPPLIERS, KM	RELIABILITY/ QUALITY
Raw materials for commercial milling:	good	350	<7km	high
Fuel	good	1	12km	good
Tools & accessories	poor	1	38	poor
Spares	poor	1	38	poor
Maintenance	average	3	<5km	average
Small repairs	average	1	38	average
Major repairs	poor	1	543	good
Transport	average	3	4km	expensive

Comments: 1.Good supply of maize for commercial milling. Customers for customs milling will sell maize at the mill.

2. Problem with spares and repair service. Some spares are available in Chipata., others in Lusaka 543km away.

3. Major overhauls only possible in Lusaka. Chipata can do exchange of basic spares. Local mechanics have inadequate knowledge and facilities.

CONCLUSIONS:

–The main demand is for dehulled and fermented meal. This will allow for the combination of traditional and mechanized milling/dehulling to ensure the best nutritional option & substantial saving in milling time.

– There is a foundation for a hammermill unit.

– Proposed model : Saro mill with 2-cylinder Kirloskar diesel engine

– Potential market for commercial milling to improve capacity. Ask for dehulled and fermented grain. Milled product will be easy to sell.

– Potential market for a dehuller driven by same engine as the grinder.

– Training needed in management at all levels.

– Training needed of diesel mechanics in Chipata and Chadiza.

– Assist supplier of spares in maintaining a better stock in Chipata.

– Look for a local trader in Chadiza to store basic spares.

Profitability Projection

Inflation makes it necessary to design the profitability projection approach around a system enabling regular adjustments of costs. For this purpose the profitability projection will be built around total income, and variable and fixed costs allowing adjustments to inflation. The income must match the costs and therefore it is the costs which should determine the prices of products/services provided. The incurred costs fall into two categories as follows.

Variable costs can be defined as costs related to the output level. More tins milled means more fuel and spare parts used.

Fixed costs are not directly related to the output level. The interest on the loans has to be paid irrespective of how much is produced. The same can be said about the watchman's salary.

The basic idea of good investment is generation of enough funds to be able to make profit and reinvest in new machinery to maintain operation and that requires realistic cost calculations.

Capital costs money, i.e. paid interest on borrowed money. Your own money has a market value too, i.e. interest earned from money saved and lent. Therefore all invested money has a cost and a common practice is to charge the present interest rate charged by banks.

Due to the high inflation rate, the costings must be adjusted each quarter. It is important that the TEAs allocate time on this matter regularly in co-operation with the Farm Management Officers. The updating will include survey on actual costs, recalculations and dissemination of findings to milling plants. A specific section is allocated for setting and changing milling charges. A common practice is to add 15 per cent on the break-even charges for the final charges.

The value of increased direct costs like fuel and spare parts are obvious but indirect costs like accumulation for future investments can be more difficult to grasp. A basic principle is to work with existing reinvestment costs and this value must be up-dated each quarter. By following the format in Case 4.5 all these aspects will be properly covered.

The intention with the costing is to be able to charge full costs for a service/product. However, competition and pressure from customers can make it difficult. A sad fact is that those charging a too low price will soon go out of business if they have no other means to subsidize a loss-making enterprise.

Most milling units will need assistance to set proper charges. The TEA cannot make individual cost calculations for all units in his district. Normally it is enough to inform about the costs for an average unit and recommend a new price list. In the end it is always the market which determines the final price.

A difficult part of profitability projections of milling projects is how to value the nutritional aspects. A good nutritious product in low demand will soon drive a miller out of business, while a less nutritious product in good demand can make good business. The former will often need sponsored campaigns to be able to compete with the commercial promotion of the latter.

CASE 4.5

PROFITABILITY PROJECTION

PROJECT: Hammermill, customs milling	
VILLAGE: Kazimuli	DISTRICT: Chadiza
OWNER: Primary Society	PERIOD: 1ST QUARTER 1992

1. ASSUMPTIONS

PRODUCTION/SALES Customs milling	QTY	UNIT	PRICE/ UNIT	TOTAL
Mealie meal	16,800	TINS	38	638,400
Bran	1,500	TINS	8	12,000
(A) TOTAL INCOME:				650,400

2. VARIABLE COSTS

Estimated basic production: **12 TINS/HOUR OR 1,400 HRS/YEAR** ITEMS	COST/ ITEM, KWA- CHA	INTER- VAL/ LIFE- SPAN, HOURS	COST/ HOUR	COST/ UNIT Tin
1. Fuel; 2.8 L/hour @ K37/L			104	8.67
2. Collection of fuel/trip	1,700	140	12	1
3. Screen	2,500	350	7	0.58
4. Oil filter	550	300	2	0.17
5. Fuel filter	650	300	2	0.17
6. Air filter	1,450	300	5	0.42
7. Oil; 8 L/change @ K300/L	2,400	300	8	0.67
8. Oil used; 2 cl per hour			6	1
9. 3 belts @ K 2,700/belt	8,100	350	23	1.92
10. Service	20,000	3,000	7	0.58
11. Set of hammers	8,000	800	10	0.83
12. Stationery			5	0.42
13. Casual workers	8,000	160	50	4.17
14. Various costs			5	0.42
(B) TOTAL VARIABLE COSTS:			246	21
1,400 HOURS/16,800 TINS/YR:			344,400	344,400

3. FIXED ANNUAL COSTS

ITEM	ACQUI- SITION COST	LIFE- SPAN	INTE- REST RATE %	ANNUAL COSTS
(C) DEPRECIATION:				
Mill	316,000	4		79,000
Initial spares set	11,000	4		2,750
Transport	9,000	4		2,250
Accessories	3,000	4		750
Shed & foundation	64,000	16		4,000
SUB-TOTAL	403,000			88,750
(D) AVERAGE CAPITAL COST (INTEREST): (based on ½ tot. acq. costs)	201,500		38	76,570
(E) SALARIES:				
Operator (full-time)				27,000
Cashier (½ time)				10,800
Watchman (½ time)				9,000
SUB-TOTAL				46,800
(F) TOTAL FIXED COSTS:				212,120

4. SUMMARY

ITEMS	TOTAL ANNUAL	TOTAL/ QUARTER
TOTAL INCOME (A)	650,400	162,600
TOTAL OPERATING EXPENSES (B) + (E)	390,528	97,632
PROFIT BEFORE INTEREST & DEPRECIATION (A) - (B) - (E)	259,872	64,968
LESS INTEREST ON CAPITAL (D)	76,570	19,142.5
PROFIT BEFORE DEPRECIATION	183,302	45,825.5
LESS DEPRECIATION (C)	88,750	22,187.5
NET PROFIT BEFORE TAX	94,552	23,638
TAX (45%)		
NET PROFIT AFTER TAX		

COST ANALYSIS

PRESENT COST LEVEL

5. TOTAL COSTS AT VARIOUS LEVELS

PRO-DUC-TION LEVEL %	PRO-DUC-TION HRS/YEAR	PRO-DUC-TION TINS/YEAR (P)	FIXED COSTS (FC)	VARIABLE COSTS (VC)	TOTAL COSTS TC = (FC+VC)
-40%	1,000	12,000	212,120	245,520	457,640
-17%	1,200	14,400	212,120	294,624	506,744
Planned	1,400	16,800	212,120	343,728	555,848
+14%	1,600	19,200	212,120	392,832	604,952
+29%	1,800	21,600	212,120	441,936	654,056
+43%	2,000	24,000	212,120	491,040	703,160

6. BREAK-EVEN ANALYSIS

PRODUCTION LEVEL %	TINS/YEAR (P)	TOTAL COSTS (TC)	BREAK-EVEN CHARGE/TIN BEC = (TC:P)	TOTAL INCOME (TI)
-40%	12,000	457,640	38.14	457,640
-17%	14,400	506,744	35.19	506,744
Planned	16,800	555,848	33.09	555,848
+14%	19,200	604,952	31.51	604,952
+29%	21,600	654,056	30.28	654,056
+43%	24,000	703,160	29.3	703,160

COST LEVEL; ONE QUARTER LATER

Note: Fill in missing data in 7 to 9 for adjusting milling charges

7. TOTAL COSTS AT VARIOUS LEVELS

PRO-DUC-TION LEVEL %	PRO-DUC-TION HRS/YEAR	PRODUC TION TINS/YEAR (P1)	FIXED COSTS (FC1)	VARIABLE COSTS (VC1)	TOTAL COSTS TC 1= (FC1+VC1)
-40%	1,000	12,000			
-17%	1,200	14,400			
Planned	1,400	16,800			
+14%	1,600	19,200			
+29%	1,800	21,600			
+43%	2,000	24,000			

8. BREAK-EVEN ANALYSIS; ONE QUARTER LATER

PRODUC-TION LEVEL %	TINS/YEAR (P1)	TOTAL COSTS (TC1)	BREAK-EVEN CHARGE/TIN BEC1 = (TC1:P1)	TOTAL INCOME (TI1)
-40%	12,000			
-17%	14,400			
Planned	16,800			
+14%	19,200			
+29%	21,600			
+43%	24,000			

9. CHANGES IN COSTS AND CHARGES

TINS/YEAR (P)	OLD BREAK-EVEN CHARGES (BEC)	CURRENT CHARGES/TIN (C/T)	NEW BREAK-EVEN CHARGE (BEC1)	NEW CHARGE PER TIN (C/T1)
12,000	38.14	38		
14,400	35.19	38		
16,800	33.09	38		
19,200	31.51	38		
21,600	30.28	38		
24,000	29.3	38		

Note: – The latest prices should be used for the quarterly adjustment.

– The investment costs for calculating the fixed costs should always

be the **latest reinvestment costs.** This would ensure adjustment for

inflation and creation of resources for replacement when machine is

worn out.

– It is important to repeat the profitability projection each quarter when

inflation goes above 25% per year. This procedure will allow for

setting proper milling charges at least 4 times per year.

Cash-flow Analysis

Case 4.6 follows the standard principles of making cash-flow analysis as well as quarterly updating due to inflation with an estimated inflation factor. In this case we have chosen 8 per cent increase per quarter (factor 1.08) which conforms with an annual inflation rate of approximately 26 per cent. Inflation rate per quarter of 26 per cent (factor 1.26) conforms with about 100 per cent annual inflation. This up-dating is necessary in order to create realistic conditions. Check the present statistics on inflation rate and determine a realistic quarterly factor. Use the following approximate formula to estimate the quarterly inflation rate from annual rate:

All processing units would benefit from proper cash-flow analyses. However, this is a rather complicated but necessary exercise particularly for new plants. The TEA must ensure that a cash-flow analysis be made available to at least each new processing plant. The first year is particularly demanding but with inflation, the financial pressure can be reduced provided there is a proper pricing system of milling charges.

$$(1 + Q/100) \smile (1 + Q/100) \smile (1 + Q/100) \smile (1 + Q/100) = (1 + A/100)$$

- Q = Quarterly inflation rate in %
- Quarterly inflation factor = $(1 + Q/100)$

- A = Annual inflation rate in %
- Annual inflation factor = $(1 + A/100)$

CASE 4.6

Table A. CASH-FLOW ANALYSIS
First quarter 1/4/92 - 30/6/92
No. of Quarters: 16

	QUARTER 1 2/92	QUARTER 2 3/92	QUARTER 3 4/92	QUARTER 4 1/93	QUARTER 5 2/93	QUARTER 6 3/93	QUARTER 7 4/93	QUARTER 8 1/94
Total income 1/1 - 30/6; 3,600 tins per quarter Base 1992 1/7 - 31/12; 4,800 tins per quarter	140,000	185,200	185,200	140,000	140,000	185,200	185,200	140,000
Inflation factor (Estim. 8% cost increase/quarter) (1.08)	1	1	1	1	1	1	2	2
Adjusted income	140,000	185,200	185,200	140,000	140,000	185,200	370,400	280,000
Principal loan	280,000							
(A) CASH INFLOW	420,000	185,200	185,200	140,000	140,000	185,200	370,400	280,000
Total operating expenses	85,350	109,900	109,900	85,350	85,350	109,900	109,900	85,350
Adjusted expenses	85,350	109,900	109,900	85,350	85,350	109,900	219,800	170,700
Project acquisition costs	403,000							
Interest on prinipal loan (38%)	26,600	26,600	26,600	23,940	21,280	18,620	15,960	13,300
Installments: Credit period; 3 years		28,000	28,000	28,000	28,000	28,000	28,000	28,000
(B) CASH OUTFLOW	514,950	136,500	164,500	137,290	134,630	156,520	263,760	212,000
(C) NET CASH FLOW (A-B)	-94,950	48,700	20,700	2,710	5,370	28,680	106,640	68,000
(D) ACCUMULATED NET CASH FLOW	-94,950	-46,250	-25,550	-22,840	-17,470	11,210	117,850	185,850

Table B. CASH-FLOW ANALYSIS
First quarter 1/4/92 - 30/6/92
No. of Quarters: 16

	QUARTER 9 2/94	QUARTER 10 3/94	QUARTER 11 4/94	QUARTER 12 1/95	QUARTER 13 2/95	QUARTER 14 3/95	QUARTER 15 4/95	QUARTER 16 1/96
Total income 1/1 - 30/6; 3,600 tins per quarter Base 1992 1/7 - 31/12; 4,800 tins per quarter	140,000	185,200	185,200	140,000	140,000	185,200	185,200	140,000
Inflation factor (Estim. 8% inflation/quarter) (1.08)	2	2	2	2	3	3	3	3
Adjusted income	280,000	370,400	370,400	280,000	420,000	555,600	555,600	420,000
Principal loan								
(A) CASH INFLOW	280,000	370,400	370,400	280,000	420,000	555,600	555,600	420,000
Total operating expenses	85,350	109,900	109,900	85,350	85,350	109,900	109,900	85,350
Adjusted expenses	157,898	219,800	237,384	198,866	215,082	298,928	323,106	270,560
Project acqusition costs								
Interest on principal loan (38%)	10,640	7,980	5,320	2,660				
Installments: Credit period; 3 years	28,000	28,000	28,000	28,000				
(B) CASH OUT FLOW	196,538	255,780	270,704	229,526	215,082	298,928	323,106	270,560
(C) NET CASH FLOW (A-B)	83,462	114,620	99,696	50,474	204,918	256,672	232,494	149,440
(D) ACCUMULATED NET CASH FLOW	269,312	383,932	483,628	534,102	739,020	995,692	1,228,186	1,377,626

GLOSSARY

Note: This is not exhaustive: some technical and scientific terms are defined within the main text, others can be found in a dictionary.

Aspiration The sorting of grain from both heavier and lighter contaminants by blowing air through it.

Bran A loose term for the whole outer covering of the grain excluding the husk. The bran frequently contains the germ as well as several layers rich in pigments, woody substances and often, anti-nutritional factors.

By-products That part of the grain which is not used for human food, but is not waste.

Chapatti A flat bread made from thinly rolled dough of wheat sorghum or other high-extraction fine flour. It is baked on a very hot surface.

Chaff The husk of wheat and similar grains which does not remain on the grain after harvest but contaminates the grain and must be removed by aspiration or winnowing.

Dehulling Is the removal of bran and the other outer layers of grain which has either already been dehusked or has been threshed naked – without the husk.

Dehusking Is the removal of husk from paddy rice and from oats - that is, from grains where the husk is still attached to the grain after threshing.

Endosperm This term is used of the grain without the husk or bran. It is the main part of the grain that is used for food and is particulary rich in starch.

Enzyme Naturally occurring substances which create change in the main chemical constituents of plant materials. For example sprouted grains contain enzymes which will convert starch to sugars in the malting process.

Extraction rate That proportion of the grain which emerges from the milling system as food.

Thus an extraction rate of 80 per cent means that 20 per cent of the grain going into the mill has been removed as by-products and 80 per cent emerges as flour or whole grain for food.

Feed Grain or other plant material used for animal feed.

Flat breads The collective name for leavened and unleavened breads that are rolled very thinly for baking, usually on a hot plate. (See chapatti.)

Flour Finely ground cereals usually passing a 150 micron mesh (0.15mm). (See meal.)

Fodder The plant material remaining after threshing or directly cut for the purpose (see feed.)

Food In the context of this book means human food.

Gari Fermented and dried cassava flour, see Chapter 2.

Gelatinization Starch particles burst open when heated with water, resulting in changes in the texture of the food, for example by it going thick and sticky.

Germ That part of the grain which will become the new plant if the grain is allowed to grow by moistening it.

Glumes The correct terms for husk and the chaff of grains: the covering which protects the grain while it is growing and ripening and may be threshed from the grain or in the case of rice and barley remain with it after threshing.

Grinding breaking and reduction in size to finer particles (see milling).

Hulls A term used loosely used for both the outer covering of grains and the husk of rice and barley, which is the correct term. Correctly, the outer bran layers of sorghum and maize.

Husk Correctly the outer covering of rice as paddy and barley. Husk is the glumes which in the case of these grains adhere to the grain after threshing (see chaff, glumes).

Leavened Fermented with yeast. As in the case of conventional bread and certain fermented foods of northern Africa. The yeast may be added by the baker or occur naturally on the grain and in its flour.

Meal Coarse flour often with the fine flour sieved from it. This is used for porridge (see semolina).

Milling The term milling is somewhat loose: with wheat and maize it means the grinding of the grain to flour or meal (coarse flour), often with some separation of the less desirable bran and germ. For rice the term means the careful removal of the husk and bran to give as high a yield of whole grain as possible. In this book, to avoid misunderstanding we shall use the term grinding when the grain is converted to meal or flour. If much of the bran is taken out, the term extraction milling is used.

Paddy Rice as it is threshed from the plant and is covered with the husk is known as paddy.

Semolina Partly milled grain of a coarse but uniform particle size.

Stover Is the name given to the straw of the thick stemmed grains sorghum, maize and pearl millet. It may be used for fodder, for thatch or fencing and other support structures.

Thatch Straw and stover which is used for roofing.

Threshing To obtain the grain from the stalk by manual, animal or mechanical processes of stripping or beating to yield the grain.

Toll The organization of cereal milling can be divided into three broad groups: home, toll (customs) or commercial. In the first two groups the ownership of the grain does not change (as in subsistence farming). Although the scale of milling will vary, the technology used is equally appropriate to all three groups. In toll milling the owner takes the grain to the mill for processing and keeps all or some of the products. The miller charges a fee for the milling, which may be in cash or in kind – for example by keeping some of the grain or its products.

Toxin Means poison. Toxins may be part of the plant material as in the case of cyanide in cassava or may be introduced by bad storage as is the case with the toxins introduced from fungus growing on damp stored grains or on grains wetted in the fields, or from residues of pesticides or other chemicals used in the farming process.

Whitening Is the removal of bran from brown rice to give white rice. The term 'polishing' is sometimes used for this process but is not the correct term.

Winnowing is the separation of grain from chaff and other contaminants by use of the wind: the threshed grain is thrown in the wind by shovels, winnowing mats or baskets.

FURTHER READING AND INFORMATION

APPENDIX 2

REFERENCES

Acland, J.D. (1971) *East African Crops,* Food and Agricultural Organization (FAO), Rome.

Alnwick, A, Moses, S. and Schmidt, O.G. (1987) *Improving young child feeding in Eastern and Southern Africa.* International Development Research Centre, Ottawa (IDRC), Canada.

Bassey, M. W. and Schmidt, O.G. (1989) *Abrasive-disk dehullers in Africa; from research to dissemination.* IDRC.

Bhat, B. A. and Uhlig, S. J. (1979) *Choice of technique in maize milling.* David Livingstone Institute, Edinburgh.

Boie, W. (1989) Introduction of animal-powered cereal mills. GATE.

Bruinsma, D.H., Witsenburg, W.W. and Wurdemann, W. (1985) Selection of technology for food processing in developing countries.

Brune, M., Rossander-Hulten, L., Hallberg, L., Gleerup, A. and Sandberg, A.S. (1992) *Phytate hydrolysis by phytate in cereals.* Dept. of Medicine II and Clinical Nutrition, University of Gothenburg, Sahlgrenska Hospital, and Dept of Food Science, Chalmers University of Technology, Gothenburg, Sweden.

Carr, M. (1989) *Women and the food cycle.* IT Publications.

Cecil, J.E. (1986) *Roller milling sorghum and millet grain using a semi-wet process.* Tropical Development and Research Institute, London, UK.

Dendy, D.A.V. (1977) Proceedings of a symposium on sorghum and millets for human food; Tropical Products Institute, London, UK.

Farm Management Handbook, (1982) Department of Agriculture, Technical and Extension Services. Zimbabwe.

Eastman, P. (1980) *An end to pounding; A new mechanical flour milling system in use in Africa.* IDRC.

Ekström, N. (1975) *Syrabehandling av spannmål.* Jordbrukstekniska Institutet, Swedish University of Agricultural Sciences.

FAO. (1989) *Utilization of tropical food: cereals.* FAO Food and Nutrition Paper 47/1. FAO, Rome.

FAO. (1984) *Rice parboiling.* FAO, Rome.

FAO. (1983) *Processing and storage of food grains by rural families.* FAO, Rome.

FAO. (1977) *World food survey.* FAO, Rome.

Henderson, S.M. and Perry, R.L. (1976) *Agricultural process engineering.* Van Nostrand Reinhold.

ILO. (1984) *Small-scale maize milling.* Technical memorandum No.7. International Labour Office (ILO), Geneva.

ILO. (1984) *Improved village technology for women`s activities.* A manual for West Africa. ILO, Geneva.

Ian Carruthers and Marc Rodrigues. (1992) *Tools for agriculture: A guide to appropriate equipment for small-scale farmers.* Intermediate Technology Publications.

International Association for Cereal Chemistry. (1976) *Sorghum and millets for human food.*

IRDC. (1980) *Nutrion och Lantbruk — Om samspelet mellan människan, jordbruket och födan. En princip diskussion med praktiska exampel från Öst Afrika.* International Rural Development Centre, Swedish University of Agricultural Sciences, Uppsala.

IRDC. Ulf Svanberg. (1992) *Fermentation of cereals: Traditional household technology with nutritional potential for young children.* International Rural Development Centre, Swedish University of Agricultural Sciences, Uppsala.

Jantzen, D.E. and Koirala, K. (1989) *Micro-hydropower in Nepal. Development effects and future prospects with special reference to the heat generator.*

Kent, N. L. (1983) *Technology of cereals* (3rd edition) Pergamon Press.

Larsson, K. (1988) *Beredning och hantering av kraftfoder — med tonvikt på foder med hög vattenhalt.* Jordbrukstekniska Institutet, Swedish University of Agricultural Sciences.

Larsson, M. and Sandberg, A.S.(1990) *Phytate reduction in bread oat flour, bran or rye bran.* Dept. of Food Science, Chalmers University of Technology, Gothenburg, Sweden.

Meire, U. (1988) Local experience with micro-hydro technology. SKAT, St. Gallen.

Metzler, R., Scheuer, H. and Yoder, R. (1984) *Small water turbine for Nepal: The Butwal experience in machine development and field installation.*

Le Ministère de la Cooperation et du Développement et le Groupe de Recherche et d'Echanges Technologiques. (1988) *Du grain à la farine.* GRET, Paris.

Ministry of Agriculture, Food and Fishery 1992. *Milling and oil extraction: A training manual for Agricultural Engineering Extension Staff.* Zambia.

Munch, L. and Pomeranz, Y. (1981) *Cereals: A renewable resource, theory and practice.* The American Association of Cereal Chemists.

Ninje, T. and Weaver, F.J. (1984) *Protein, energy and carotene content of maize at six stages in the preparation of 'ufa' by village processing methods.* Ecology of Food and Nutrition vol.15; pp. 237-280.

Nojonen, M.A. (1989) *Lantbrukets arbetsmiljö.* Institutet för Arbetshygien, Helsingfors, Finland.

Noren, O. (1990) *Rapsolja för tekniska ändamål — framställning och användning.* Jordbrukstekniska Institutet, Swedish University of Agricultural Sciences, Uppsala.

Passmore, R. and Eastwood, M. A. (1986) *Human nutrition and dietetics* (8th edition) Churchill Livingstone, Edinburgh.

Pedersen, T. T. (1966) *Kornkernens slagstyrkeslagemöllens effektbehov.* Den KGL. Veterinaer og Landbrugshögskole, Afdelingen för Landbrugsmaskiner.

Pomeranz, Y. (1988) *Wheat: Chemistry and Technology.* American Association of Cereal Chemists.

The Rural Solar Mill. (1992) 11th European Photovoltaic Conference and Exhibition, Spain.

Sandberg, A.S. (1991) *The effect of food processing on phytate hydrolysis and availability of iron and zinc.* Dept. of Food Science, Chalmers University of Technology, Gothenburg, Sweden.

Sandberg, A.S. and Svanberg, U. (1991) *Phytate hydrolysis by phytase in cereals; Effects on in-vitro estimation of iron availability.* Dept. of Food Science, Chalmers University of Technology, Gothenburg, Sweden.

SKAT. (1986) *Evaluation et choix de moulins.* Swiss Centre for Appropriate Tchnology, St. Gallen.

Steckle, J. (1974) *Consumer preference study in grain utilization,* Maiduguri, Nigeria, Food and Nutrition Sciences Division, African Regional Office, Dakar, Senegal.

UNIFEM/WAFT. (1986), *Cereal processing.* Food Cycle Technology Source Book No.3.

UNU. (1984) *Interfaces between agriculture, nutrition and food science.* The United Nation University, India.

Vanek, K.V. (1986) *Small-scale grain processing schemes in Africa: Methods, equipment, energy and power requirement use of renewable energy resources.*

VIS. (1992) *Hammer mill Operator's Manual.* Lusaka, Zambia (Draft).

VIS & ZAMS. (1991) *Preliminary Hammer mill Survey.* Village Industrial Services and Zambian Agribusiness Support project.

Wimberly J. E. (1983) *Paddy-rice postharvest industry in developing countries;* International Rice Research Institute, Manila.

Yde, L. (1988) *Vindkvaern till U-Lande.* Nordvestjysk Folkecenter for Vedvarende Energi.

ZAMS. (1991) *Economics of the hammer mill operation in Northern and Eastern provinces of Zambia.*

INSTITUTES AND RESEARCH BODIES WITH INTERESTS IN SMALL-SCALE MILLING

AFRICA

Centre Horticole et Nutritionnel de Ouando
BP 13, Porto Novo, Benin

Ministry of Agriculture
P.O.Box Gaborone, Botswana

Rural Industries Innovation Centre (RIIC)
Private Bag 11, Kanye, Botswana

I.R.A.T.
BP 7047, Bobo Dioulasso, Burkina Faso

Institut de Recherches en Biologie et Ecologie Tropical
BP 7047, Ouagadougu, Burkina Faso

I.I.T.A.
BP 1495, Ouagadougu, Burkina Faso

Service de la Technologie, s/c Nutrition,
Direction General de Alimentaire et de l'Agriculture
BP 7028, Ouagadougu, Burkina Faso

West African Rice Development Association (WARDA)
01 BP 2551, Bouake,Côte d'Ivoire

Societé Ivoirienne de Technologie Tropicale
04 BP 1137, Abijan, Côte d'Ivoire

Institute of Agricultural Research
PO Box 3001, Addis Ababa, Ethiopia

International Livestock Centre for Africa
PO Box 5689, Addis Ababa, Ethiopia

National Cereal Research Institute
PO Box 5654, Addis Ababa, Ethiopia

Sorghum Improvement Programme
Institute of Agriculture Research, Nazret Research Station
PO Box 103, Nazret, Ethiopia

Institute of Development Research,
Addis Ababa University
PO Box 1176, Addis Ababa, Ethiopia

Catholic Relief Services
PO Box 569, Banjul, Gambia

Food Research Institute
PO Box M20, Accra, Ghana

University of Science and Technology,
Faculty of Agriculture, Department of Engineering,
Kumasi, Ghana

Kenya Industrial Research and Development Institute
PO Box 30650, Nairobi, Kenya

Department of Food Science and Technology,
University of Nairobi
PO Box 29053, Kabete, Kenya

Nutrition Centre
PO Box 557, Mafeteng, Lesotho

Division of Agricultural Research, Min. of Agriculture
PO Box 829, Maseru, Lesotho

West African Rice Development Associotion (WARDA)
PO Box 1019, Monrovia, Liberia

Chitedze Agricultural Research Station,
Farm Machinery Unit
PO Box 158, Lilongwe, Malawi

Office Produits Agricoles du Mali (OPAM)
B P 132, Bamako, Mali

CEEMA, Division du Machinisme Agricole,
Directionale de Genie Rural, Ministere de l'Agriculture
BP 155, Bamako, Mali

Institut d'Economie Rurale
Rue Mohamed V, Bamako, Mali,

S.R.C.V.O. (Ministere de l'Agriculture)
BP 438, Bamako, Mali

C.M.D.T.
BP 487,Bamako, Mali

Instituto National de Investigaciones Agronomica
P 3658, Mavalane-Maputo 11, Mozambique

Inst. of Study & Application of Integrated Development
BP 2821, Nimery, Niger

I.N.R.A.N.
BP 429, Niamey, Niger

C.N.R.A. Tarna
BP 240, Maradi, Niger

Federal Institute of Industrial Research,
PMB 1023, Oshodi, Ikeya, Lagos, Nigeria

International Institute for Tropical Agriculture (IITA)
PMB 5320, Ibadan, Nigeria

Agricultural Development Authority
Enugu, Anambra State, Nigeria

Root Crop Research Institute
Umudike, Umuahla, Imo State, Nigeria

National Cereals Research Institute (NCRI),
Badeggi, PMB 8 Bida, Niger State, Nigeria

Department of Food Science and Technology,
University of Ife, Ile-Ife, Nigeria

National Food Research Institute
PO Box 395, Pretoria 0001, Republic of South Africa

Institut de Tecnologie Alimentaire
BP 2765, Dakar, Senegal

Societe Industrielle Sahelienne de mecanique
de materiel agricoles et de representations
BP 3214, Dakar, Senegal

C.R.D.I
BP 1107, Dakar, Senegal

Societe Nouvelle du Development
Rue Bayeux, Dakar, Senegal

Societe de Development et de Vulgarisation Agricole
BP 3234, Dakar, Senegal

C.E.G.I.R. (Etude de Marketing)
Avenue Bourghiba, Dakar, Senegal

Institut Senegalaise de Recherche Agronomique (ISRA)
BP 53, Bambey, Senegal

Louis Berger International
44 Rue Jules Ferry, Dakar, Senegal

Tikonko Agricultural Extension Centre
PO Box 142, BO, Sierra Leone

Food Research Centre
PO Box 213, Khartoum, Sudan

Tanzania Food and Nutrition Centre
PO Box 977, Dar es Salam, Tanzania

Small Industries Development Organisation
PO Box 2476, Dar es Salam, Tanzania

Department of Food Science and Technology
Sokoine University of Agriculture
PO Box 3006, Morogoro, Tanzania

Community Development Trust Fund
PO Box 9421, Dar es Salam, Tanzania

Centre for Agric. Mechanization and Rural Technology
PO Box 764, Arusha, Tanzania

Institut National de Plantes a Tubercules
BP 4402, Lome, Togo

Crop Storage and Processing Unit,
National Research Council
PO Box 6884, Kampala, Uganda

Serere Agricultural Research Station
PO Soroti, Uganda

CEDECO
BP 70, Kimpese, Zaire

National Council for Scientific Research
PO Box CH 158, Lusaka, Zambia

TDAU, University of Zambia
PO Box 32379, Lusaka, Zambia

Northern Technical College
Ndola, Zambia

Small Industries Development Organisation
PO Box 35373, Lusaka, Zambia

Village Industrial Service
PO Box 35500, Lusaka, Zambia

Food Science Department, University of Zimbabwe
PO Box MP45, Harare, Zimbabwe

AGROTEC
PO Box W 540, Harare, Zimbabwe

Nutrition Section, Ministry of Health
Harare, Zimbabwe

SADCC/ICRISAT
PO Box 776, Bulawayo, Zimbabwe

Silveira House
PO Box 545, Harare, Zimbabwe

Environment Development Activities-Zimbabwe
PO Box 3492, Harare, Zimbabwe

ASIA

Bangladesh Rice Research Institute (BRRI)
PO Box 911, Ramma, Dacca, Bangladesh

Post-harvest Technology Application Centre (PTAC)
70 Pansodan St., Burma

Central Food Technological Research Institute (CFTRI)
Chelvamba Mansion, V.V. Mohala, PO Mysore 13,
India

Internatational Crop Research Institute for
the Semi-Arid Tropics (ICRISAT)
Patancheru PO, Hyderabad 502, 324, A.P. India

Paddy Processing Research Centre
Kheragpur 721302, W Bengal, India

Protein Foods and Nutrition Development Association
of India, Bombay, India

Agricultural Research Institute
New Delhi, India

Central Research Institute for Agriculture
Bogor, Indonesia

IRRI-Pak Agricultural Machinery Programme
Islamabad, Pakistan

Pakistan Agricultural Research Council
Islamabad, Pakistan

National Grain Authority
Quezon City, Philippines

International Rice Research Institute (IRRI)
PO Box 933, Los Banos, Philippines

Philippines Council for Agriculture and Resources
Research, Los Banos, Laguna, Philippines

Singapore Council of Scientific and Industrial Research
(SISIR), 179 River Valley Road, Singapore 0617

Ceylon Inst. of Scientific and Industrial Research (CISIR)
363 Baudhaloka Mawatha, Colombo 7, Sri Lanka

Rice Processing Research and Development Centre
Jayanthi Mawatha, Stage 2, Anuradhapura, Sri Lanka

International Crops Research Institute for
Semi-Arid Tropics (ICRISAT), Aleppo, Syria

Asia Institute of Technology
PO Box 2754, Bangkok, Thailand

Thailand Inst. for Scientific and Technological Research
Bangkok, Thailand

AUSTRALIA

Wheat Research Institute
Toowoomba, Australia

Yanco Agricultural Institute
2703 Yanco, NSW, Australia

NORTH AMERICA

International Development Research Centre (IDRC)
PO Box 8500, Ottawa, Canada

The Plant Bio-Technology Institute, National Research
Council of Canada 110 Gymnasium Raod, Saskatoon
Saskatschewan, Canada S7 OW9

Canada Hunger Foundation
75 Sparks St., Ottawa, Canada

University of Guelp, Department of Food Science
Guelp, Ontario, Canada NIG ZWI

University of Saskatschewan, Department of Crop
Science, Saskatoon, Saskatschewan, Canada 75N OWO

Department of Foods and Nutrition, Kansas State
University, Justin Hall Manhattan, Kansas 66506,
USA

Cereal Quality College Station, Laboratory Department
of Soil & Crop Science
Texas A & M University, Texas 77843, USA

Volunteers in Technical Assistance (VITA)
1815 North Lynn St, Suite 200,
Arlington, Virginia 22209, USA

American Association of Cereal Chemists
(AACC)   3340, Pilot Knob Road, St. Paul,
Minnesota, USA

International Food Policy Research Institute
1776 Massachusetts Avenue NW, Washington D.C.
20036, USA

Western Regional Research Centre
Berkeley, California 94710, USA

Southern Regional Research Centre
New Orleans, Louisiana, USA

Northern Regional Research Centre
Peoria, Illinois 61604, USA

Grain Marketing Research Laboratory
1515 College Ave, Manhattan, Kansas 66502, USA

INTSORMIL, 241 Keim Hall, University of
Nebraska, Lincoln, Nebraska 68583-0723, USA

CENTRAL AND SOUTH AMERICA

Instituto de Technologica Alimentos
Caixa Postal 139, 13.100 Campinas, SP, Brasil

Fundaoao Centro Technologico de Minas Gerais
Minas Gerais, Brazil

Centro Internacional de Agricultura Tropical
Palmira, Cali, Colombia

Instituto de Investigaciones Technologicas
Bogota, Colombia

Centro de Investigaciones y Technologia Alimentos
Universidad de Costa Rica, San José, Costa Rica

Instituto Superior de Agric
Aptdo 166, Santiago, Dominican Republic

Centro de Desarrollo Industrial de Ecuador Guayaquil,
Ecuador

Institute for Nutrition in Central America and Panama
Guatemala City, Guatemala

Caribbean Food and Nutrition Institute
Kingston, Jamaica

Centro Internacional para Maiz y Trigo
El Batan, Mexico

Laboratorio de Technologia de Alimentos,
Min de Industria, Aptdo 189, Managua, Nicaragua

Instituto de Investigaciones Tropical
La Molina, Aptdo 11294, Lima 14, Peru

Ministry of Agriculture Lands and Fisheries
Port-of-Spain, Trinidad and Tobago

EUROPE

Intl. Association for Cereal Science & Technology
A2320, Schwechat, P.O.Box 77, Wieneerstrasse 22a,
Vienna, Austria

Universite Catholique de Louvain, Faculte de Sciences
Agronomiques, 3 Place Croix du Sud, 1348 Louvain La
Neuve, Belgique

Bundesforschunganstalt fur Getreideverarbeitung
D-4930 Delmold 1, Germany

Germany Agency for Technical Cooperation (GTZ)
BP 5180, D- 6236 Eschborn, Germany

German Appropriate Technology Exchange (GATE)
Postfach 5180, D-6236 Eschborn, Germany

Projekt-Consult Beratung in Entwicklungsländern
GmbH Limburger StraBe 28, D-6240, Königstein,
Germany

ALTERSIAL (ENSIAA)
1 Avenue des Olympiades 91305, Massey, France

C.I.R.A.D./C.E.E.M.A.T.
Stockage et Conservation des cereales, Decoticage du
Riz, Domain de la Valatte, 34100, Montpellier, France

I.R.A.T., Laboratoire de Technologie de Cereales
9 Place Pierre Viala, 34060 Montpellier Cedex 01,
France

MARCONBER (Etude de Marketing)
1 Rue Therese, 75001, Paris, France

Groupe de Recherche et d'Exhanges Technologique
213 Rue Lafayette, Paris 75010, France

Institut Technologique Dello
8 Rue Paul Bert, F-83300, Aubervilliers,
France

Institut National de Recherches Agronomiques
Montpellier, France

Instituut Voor Graan, Meel en Brood
Postbus 15, 6700 AA Wageningen, The Netherlands

Royal Tropical Institute
Mauritskade 63, 1092 A D, Amsterdam, The Netherlands

Institute for Cereals, Flour and Baking
TNO - Lawickse Alle 15, Wageningen, The Netherlands

Technische Hogeschool
Eindhoven, Netherlands

Flour Milling and Baking Research Association
Chorleywood, Rickmansworth, Herts, UK

Institute of Development Studies, University of Sussex
Brighton, BN1 9RE, UK

Tropical Products Institute
Cullham, Allindon- Oxon, OX14 3DA, UK

Intermediate Technology Development Group
Myson House, Railway Terrace, Rugby CV21 3BT, UK

Overseas Development and Natural Resources Institute
56-62 Gray's Inn Road, London WC1X 8LU, UK

National Institute of Agricultural Engineering
Silsoe, Bedfordshire, UK

Natural Resources Institute (NRI)
Central Avenue, Chatham, UK

Swedish Institute of Agricultural Engineering
Box 70033, 750 07 Uppsala, Sweden

Contact Forum for Technique in Developing Countries
Box 7033, 750 07, Uppsala, Sweden

Chalmers University of Technology, Dept. of Food
Science, Gothenburg, Sweden.

PRODUCT DETAILS

MACHINERY SUPPLIERS AND MANUFACTURERS

Small-scale Milling Machinery

This study depended on the co-operation of companies producing or selling small-scale milling equipment. Their assistance is sincerely appreciated. None of the companies cited are in any way responsible for anything that is written in the text.

The following is designed to give a general view of small-scale milling equipment. It is by no means exhaustive and has been made possible by the response of the manufacturers listed to the request for information. Details of the machine capabilities are for general guide purposes only and will be affected by grain quality. Where an illustration is given, this may represent only one range of available equipment, particularly where the output spans a large range.

RICE MILLING EQUIPMENT

HUSK REMOVAL

Alvan Blanch – Development Co Ltd
Chelworth, Malmesbury, Wiltshire SN16 9SG
United Kingdom

Type:	Steel Huller (manual)	Steel Huller	Rubber Roll
Output (kg/h):	15	250 – 350	1,000
Horsepower:		12 – 15	

CeCoCo Chuo Boeki Goshi Kaisha
PO Box 8, Ibaraki City, Osaka 567
Japan

Type:	Centrifugal (manual)	Rubber Roll
Output (kg/h):	250	360 – 1,200
Horsepower:		0.5 – 5

Colombini Sergio & Co C s.n.c.
Foxhills Industrial Estate, Scunthorpe,
South Humberside DN15 8QW
United Kingdom

Type:	Steel Huller (manual)	Steel Huller
Output (kg/h):	14	160 – 350
Horsepower:		12 – 15

Dandekar Machine Works
Bhiwandi – 431 302, Dist – Thane, Maharahtra
India

Type:	Rubber Roll
Output (kg/h):	900 – 1,100
Horsepower:	3 – 5

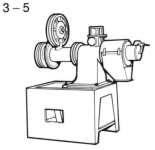

Indústrias Máquina D'Andréa S/A
Rua General Jardim, 645, 01223 – São Paulo
Brasil

Type:	Integrated huller/polisher available

Baldeschi & Sandreani s.r.l.
Lungotevere Flaminio 22, 00196 Roma
Italy

Type:	Steel Huller (manual)	Steel Huller
Output (kg/h):		110 – 1,250
Horsepower:		3 – 25

Kisan Krishi Yantra Udyog
64 Moti Bhawan, Collector Ganj,
Kanpur – 208 001
India

Type:	Centrifugal	Roller Roll
Output (kg/h):	400 – 500	1,500 – 2,000
Horsepower:	2	10

RICE MILLING EQUIPMENT

HUSK REMOVAL

Lewis C Grant Ltd
  East Quality Street, Dysart, Kirkcaldy,
  Fife KY1 2UA
  United Kingdom
Type: Steel Huller

Output (kg/h):   140 – 300
Horsepower: 12 – 15

Nogueira S/A Máquinas Agrícolas
  Rua 15 de Novembro, 781, Itapira SP 13970
  Brasil

Type: Integrated huller/polisher available

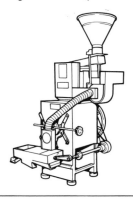

Wm McKinnon & Co Ltd
  Spring Garden Iron Works, Aberdeen AB9 1DU
  United Kingdom

Type: Steel Huller

Output (kg/h):   40 – 310
Horsepower: 5.5 – 16

Seacom, Sea Commercial Co Ltd
  PO Box SM – 202 Manila
  Philippines

Type: Integrated huller/polisher available

Satare Engineering
  Ueno Hirokohi Building
  Ueno 1 – 19 – 10, Taito – Ku
  Tokyo, Japan

Type: Integrated huller/polisher available

RICE MILLING EQUIPMENT

BRAN REMOVAL

CeCoCo Chuo Boeki Goshi Kaisha
  PO Box 8, Ibaraki City, Osaka 567
  Japan

Type:	Manual	Abrasive Cylinder
Output (kg/h):	10 – 15	300 – 1,200
Horsepower:		5 – 30

Kisan Krishi Yantra Udyog
  64 Moti Bhawan, Collector Ganj,
  Kanpur – 208 001
  India

Type:	Abrasive Cone	Abrasive Cylinder	Abrasive Disc
Output (kg/h):	400 – 1,600	400 – 700	200 – 300
Horsepower:	5 – 20	7.5 – 15	3

Dandekar Machine Works
  Bhiwandi – 431 302, Dist – Thane, Maharahtra
  India

Type:	Abrasive Cone	Friction Jet
Output (kg/h):	1,200 – 1,700	1,200 – 1,700
Horsepower:	6 – 15	15

Lewis C Grant Ltd
  East Quality Street, Dysart, Kirkcaldy,
  Fife KY1 2UA
  United Kingdom

Type:	Steel Huller
Output (kg/h):	270 – 590
Horsepower:	12 – 15

Wm McKinnon & Co Ltd
  Spring Garden Iron Works
  Aberdeen AB9 1DU
  United Kingdom

Type:	Steel Huller
Output (kg/h):	270 – 590
Horsepower:	9 – 14

STONE MILLS

ABC Hansen A/S
  Kirkegade 1
  PO Box 73, DK-8900 Randers
  Denmark

Output (kg/h): 100 – 1,000
Electric
Horsepower: 3 – 25

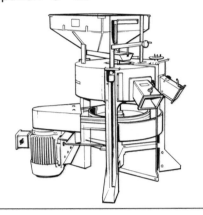

Alvan Blanch Development Co Ltd
  Chelworth
  Malmesbury
  Wiltshire SN16 9SG
  United Kingdom

Output (kg/h): 150 – 1,000
Electric/Diesel
Horsepower: 7.5 – 18

Dandekar Machine Works
  Bhiwandi – 431 302
  Dist – Thane
  Maharashtra
  India

Output (kg/h): 200 – 300
Diesel
Horsepower: 7 – 10

A/S Maskinfabrikken SKIOLD Saeby
  Kjeldgaardsvej
  PO Box 143
  DK 9300 Saeby, Denmark

Output (kg/h): 250 – 1,000
Electric/Diesel
Horsepower: 5.5 – 15

Ets Guy Moulis constructeur
  Avenue de Castres
  81360 Montredon-Labessonie
  France

Output (kg/h): 50 – 800
Electric/Diesel/Petrol
Horsepower: 1 – 6

PLATE MILLS

ABC Hansen A/C
  Kirkegade 1, PO Boz 73
  DK-8900 Randers, Denmark

Output (kg/h): 6 – 18
Electric/Petrol/Hand
Horsepower: 0.5 – 1
Artificial millstones available

Christy Hunt (Agricultural Ltd)
  Foxhills Industrial Estate
  Scunthorpe, South Humberside DN15 8QW
  United Kingdom

Output (kg/h): 7 – 275
Electric/Diesel/Petrol/Hand
Horsepower: 1 – 8

Ndume Ltd
  PO Box 62
  Gilgil
  Kenya

Output (kg/h): 15
Hand

President
  Springstrup, Box 20
  DK-4300 Holback
  Denmark

Output (kg/h): 250 – 1,100
Electric/Diesel/Hand
Horsepower: 3 – 15

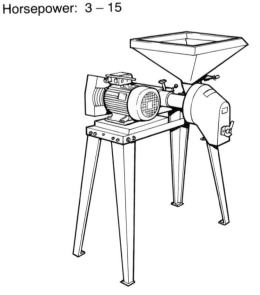

Alvan Blanch
  Chelworth, Malmesbury
  Wiltshire SN16 9SG, United Kingdom

Output (kg/h): 15 – 400
Electric/Diesel/Petrol/Hand
Horsepower: 0.5 – 10
Artificial millstones available

Cormall
  Tornholm 3
  DK-6400 Sonderborg
  Denmark

Output (kg/h): 250 – 350
Electric/Petrol/Hand
Horsepower: 5.5
Artificial millstones available

Penagas Hermanos & Cia Ltda
  Calle 28, No 20 – 80
  Apartado Aereo 689
  Bucaramanga, Colombia

Output (kg/h): 200 – 400
Electric/Diesel/Petrol
Horsepower: 3 – 6

A/S Maskinfabrikken SKIOLD Saeby
  Kjeldgaardsvej, PO Box 143
  DK-9300 Saeby, Denmark

Output (kg/h): 150 – 350
Electric/Diesel/Petrol/Hand
Horsepower: 5.5 – 11
Artificial millstones available

HAMMERMILLS

Manik Engineers Ltd
PO Box 1274, Arusha
Tanzania

Output (kg/h): 90 – 816
Electric/Diesel/Tractor (PTO)
Horsepower: 8 – 30

Philco Dierings Ltd
Forest Vale Industrial Estate, Cinderford
Gloucestershire GL14 2HP
United Kingdom

Output (kg/h): 317 – 907
Electric/Diesel/Petrol
Horsepower: 15 – 20

Mio Osijek
Vukovarska cesta 219a, Osijek
Yugoslavia

Output (kg/h): 50 – 200
Electric
Horsepower: 0.7

President
Springstrup, Box 20, DK-4300 Holbaek
Denmark

Output (kg/h): 90 – 1,500
Electric/Diesel/Petrol
Horsepower: 7.5 – 15

Ndume Ltd
PO Box 62, Gilgil
Kenya

Output (kg/h): 190 – 1,270
Electric/Diesel/Tractor (PTO)
Horsepower: 8 – 25

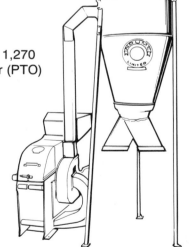

A/S Maskinfabrikken SKIOLD Saeby
Kjeldgaardsvej, PO Box 143, DK-9300 Saeby
Denmark

Output (kg/h): 80 – 1,350
Electric/Diesel/Tractor (PTO)
Horsepower: 7.5 – 80

Tradepoint
Handelsgesellschaft m.b.H., Geylinggasse 30,
A – 1130 Vienna
Austria

Output (kg/h): 800 – 1,000*

* Complete Containerized Milling System

Penagos Hermanos & Cia Ltda
Calle 28, No 20–80, Apartado Aereo 689,
Bucaramanga
Colombia

Output (kg/h): 90 – 1,500
Electric/Diesel
Horsepower: 3 – 8

HAMMERMILLS

Alvan Blanch
 Chelworth, Malmesbury, Wiltshire SN16 9SG
 United Kingdom

Output (kg/h): 110 – 1,250
Electric/Diesel/Petrol/Tractor (PTO)
Horsepower: 3 – 25

Ateliers Albert
 Rue Riverre, 4, 5750 Floreffe (Namur)
 Belgium

Output (kg/h): up to 1,500
Electric/Diesel
Horsepower: 20

Baldeschi & Sandreani s.r.l.
 Lungotevere Flaminio 22, 00196 Roma
 Italy

Output (kg/h): 150 – 1,700
Electric/Diesel
Horsepower: 10 – 30

C S Bell
 170 W Davis St., PO Box 291, Tiffin, Ohio 44883
 USA

Output (kg/h): up to 680
Electric/Diesel/Petrol
Horsepower: 2 – 19

Christy Hunt (Agrictural) Ltd
 Foxhills Industrial Estate, Scunthorpe
 South Humberside DN15 8QW
 United Kingdom

Output (kg/h): 64 – 1,500
Electric/Diesel/Petrol
Horsepower: 3 – 20

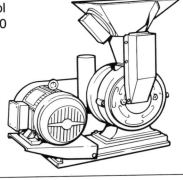

Cormall
 Tornholm 3, DK-6400 Sonderborg
 Denmark

Output (kg/h): 85 – 505
Electric/Diesel/Petrol
Horsepower: 7.5 – 12

Electra
 47170 Poudenas, Mezin
 France

Output (kg/h): 80 – 800
Electric/Diesel/Petrol/Tractor (PTO)
Horsepower: 2 – 14

Gosling Group
 Knightsdale Road, Ipswich, Suffolk IP1 4LE
 United Kingdom

Output (kg/h): 35 – 1,800
Electric/Diesel/Petrol/Tractor (PTO)
Horsepower: 5 – 40

Indústrias Máquina D'Andréa S/A
 Rua General Jardim, 645, 01233 – São Paulo
 Brasil

Output (kg/h): 80 – 1,500

Law-Denis
 Lavenham Road, Beeches Industrial Estate, Yate
 Bristol BS17 5QX
 United Kingdom

Output (kg/h): 270 – 1,250
Electric
Horsepower: 7.5 – 15

ROLLER MILLS

Baldeshi & Sandreani s.r.l. Lungotevere Flaminio 22 00196 Roma Italy Output (kg/h): 400 – 600 Electric Horsepower: 12 – 15	**T W Barfoot** Barfoot Centre Hackhurst Industrial Estate Lower Dicker, Hailsham East Sussex BN27 4BW United Kingdom Output (kg/h): 1,000 – 12,000* Electric Horsepower: 60
Favini Impianti s.r.i. Via Marconi, 13 24040 Fornovo S.G. Bergamo Italy Output (kg/h): 600 – 1,000** Electric Horsepower: 120	**Maize Master Roller Mills** Box 430 Kronstadt, OFS 9500 Republic of South Africa Output (kg/h): up to 500 Electric Horsepower: 10
Ocrim SpA Via Massarotti 76 Cremona Italy Output (kg/h): 500 – 583 Electric/Diesel Horsepower: 70 – 120	
* Container Mill ** Mobile Milling Unit/Container Mill	

MAIZE SHELLERS

Alvan Blanch Develop Co Ltd
Chelworth, Malmesbury
Wiltshire SN16 9SG
United Kingdom

Output (kg/h): 50 – 1,500
Electric/Diesel/Petrol/Tractor PTO/Manual
Horsepower: 0.5 – 7

Columbi Sergio & C s.n.c.
Via Cadorna, 9
PO Box 19, 20081 Abbiategrasso
Milano, Italy

Output (kg/h): up to 1,200
Diesel/Manual
Horsepower: 10

Electra
47170 Poudenas
Mezin
France

Output (kg/h): 10 – 20
Manual

Hira International
Hospital Road
Jagraon – 14026
Dist Ludhiana, India

Output (kg/h): 20 – 250
Electric/Diesel/Petrol/Manual
Horsepower: 3 – 5

Ateliers Albert & Co Ltd
Rue Riverre, 4
5750 Floreffe (Namur)
Belgium

Output (kg/h): 250 – 1,500
Electric/Diesel/Petrol/Tractor PTO/Manual
Horsepower: 3

CeCoCo Chuo Boeki Goshi Kaisha
PO Box 8
Ibaraki City, Osaka 567
Japan

Output (kg/h): 100 – 1,125
Manual**
Horsepower: 0.5 – 2

Christy Hunt (Agricultural) Ltd
Foxhills Industrial Estate, Scunthorpe
South Humberside DN15 8QW
United Kingdom

Output (kg/h): 75 – 750
Electric/Petrol/Manual**
Horsepower: 0.5 – 3

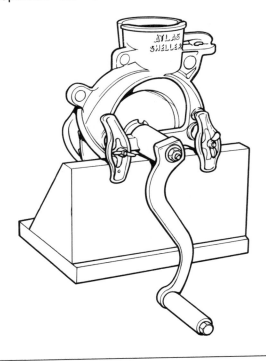

** Both hand and foot operated machines available

MAIZE SHELLERS

Kaneko Agricultural Machinery
26-11 Higashi Nihonbashi 2-chrome
Chuo-Ku, Tokyo
103 Japan

Output (kg/h): 90 – 100
Electric/Manual**
Horsepower: 0.5 – 2

Mio Osijek
Vukovarska cesta 219a
Osijek
Yugoslavia

Output (kg/h): 200 – 300*
Electric/Manual
Horsepower: 0.7 – 3

Nogueira S/A Máquinas Agrícolas
Rua 15 de Novembro, 781
Itapira SP 13970
Brasil

Output (kg/h): 900 – 2,100
Electric/Diesel/Petrol/Tractor PTO
Horsepower: 5 – 8

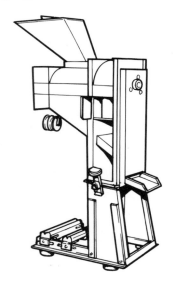

Lewis C Grant Ltd
East Quality Street
Dysart, Kirkcaldy
Fife KY1 2UA, Scotland

Output (kg/h): 180
Electric
Horsepower: 15 – 20

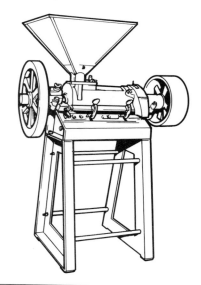

Ndume Ltd
PO Box 62
Gilgil
Kenya

Manual

Penagos Hermanos & Cia Ltd
Calle 28, No 20-80
Apartado Aereo 689, Bucaramanga
Colombia

Output (kg/h): 300 – 1,600
Electric/Diesel/Petrol
Horsepower: 0.25 – 7.5

Seedburo Equipment Company
International Division
1022 W Jackson Blvd, Chicago
Illinois 60607-2990
USA

Manual

** Both hand and foot operated machines available

* Integrated hammermill/corn sheller available

FURTHER LIST OF MANUFACTURERS AND SUPPLIERS

AFRICA

Rural Industries Innovation Centre (RIIC)
Private Bag 11, Kanye, Botswana

Granges Moulins Burkinabe
Faso-Mougu, BP 64, Banfora, Ouagadougu, Burkina
Faso

I.B.E./I.T.Dello
BP 3370, Ouagadougu, Burkina Faso

OF.NA.CER. Office National des Cereales
Ouagadougu, Burkina Faso

C.I.L.S.S.
BP 7049, Ouagadougu, Burkina Faso

Service de Coordination des O.N.G
Avenue Coulibaly, Ouagadougu, Burkina Faso

S.A.C.M.
16 Rue des Foreurs, BP 4019, Côte d'Ivoire

Catholic Relief Service (C.R.S)
3 Marina Parade, PO BOX 569, Banjul, Gambia

Agricultural Engineers Ltd
Ring Road West Industrial Area, PO BOX 3707, Accra,
Ghana

Technology Consultancy Centre, University of Science
and Technology
University Post Office, Kumasi, Ghana

Agro Machine Ltd, Liberia Industrial Free Zone
PO Mail Bag 9047, Liberia

Agrimal (Malawi) Ltd
PO BOX 143, Blantyre, Malawi

Brown and Clapperton Ltd
PO BOX 52, Blantyre, Malawi

S.R.C.V.O (Voir Organismes de Recherche)
BP 1626, Bamako, Mali

Grands Moulins du Mali
Koulikoro SOMABIPAL, Bamako, Mali

Office des Produits Agricole du Mali
BP 132, Bamako, Mali

C.M.D.T.
BP 487, Bamako, Mali

Division du Machine Agricole
BP 155, Bamako, Mali

C.C.A.
3eme arrondissement, Bamako, Mali

EURO ACTION ACCORD
Quartier Quazambouzou, Bamako, Mali

SOMATRIL
Zinder, Niger

Fabrication Engineering and Production Company,
Projects Development Institute
3 Independance Layout, PO BOX 609, Enugu, Nigeria

Grands Moulins SENTENAC
Route de Rufisque, BP 451, Dakar, Senegal

Grands Moulins de Dakar
Mole 8, Dakar, Senegal

SISCOMA
B.P., Dakar, SENEGAL

Centre National de Recherche Forestiere
BP 2765, Hann-Dakar, Senegal

Fond d'Equipement des Nations Unies, s/c Direction du
bien-être famial
BP 414, Rue Leblanc, Dakar, Senegal

Commissariat a la Securite Alimentaire
Rue Blanchet, Dakar, Senegal

Societe National de Distribution
BP 2048, Dakar, Senegal

SODEFITEX
BP 3216, Dakar, Senegal

Environment et Development Africain, Reseau
Technologique Pour le Development
BP 3370, Dakar, Senegal
D. Seck
Kebemer, Senegal

C.Gueye
Gossas, Senegal

Tikonko Agricultural Extension Centre
PO BOX 142, BO, Sierra Leone

Ubungo Farm Implements
PO BOX 2669, Dar es Salam, Tanzania

United Engineering Workers
PO BOX 3082, Arusha, Tanzania

Kaleya Engineering Ltd
PO BOX 71640, Mazabuka, Zambia

Saro Agricural Equipment Ltd
PO BOX 35168, Lusaka, Zambia

Re-United Engineering Ltd
PO BOX 30541, Lusaka, Zambia

GAMECO
PO BOX 30541, Lusaka, Zambia

Turning & Metals
Lusaka, Zambia

ENDA-ZIMBABWE
PO BOX 3492, Harare, Zimbabwe

ASIA

Allied Trading Company
Railway Road, Ambala City 134 002, Maryana, India

Cossul & Co. PVT Ltd
Industrial Area, Fazalgury, Kanpur, India

Kumaon Agri-Horticulture Stores
P.O.Kashipur, Distr Nainital, U.P., India

Mohan Singh Harbhajan Singh
G.T. Road, Goraya 144 409, Distr. Jullundur, India

Mohinder & Co. Allied Industries
Kurali, Distr Ropar, Punjab, India

Rajan Traging Co.
PO BOX 250, Madras, 600 001, India

Rajasthan State Agro Industries Corparations Ltd
Virat Bhawan, C.Scheme, Jaipur 302006, Rajasthan, India

Guanko Iron Works
102-104 Mj Cuenco Avenue, Cebu City, Philippines

AUSTRALIA

Melanesian Council of Churches
PO BOX 80, Lae, Papua New Guinea

NORTH AMERICA

Jacobson International Inc
2445 Nevada Avenue North, Minneapolis, Minnesota 55427, USA

Rait Manufacturing Co
Shenandoah, Ohio, USA

CENTRAL AND SOUTH AMERICA

Irmaos Nogueira SA & CIMAG Ltda
Av. Ipiranga 1071, São Paulo, Brazil

Laredo S.A.
Rua 1 de Agosto, 11-67 CEP, 17100 Bauru (SP), Brasil

EUROPE

A T O L
Blijde Inkamstraat 9, 300, Leuven, Belgium

DDD President
Chaussee de Dikkebus 487, 8904 Ypres, Belgium

A/S Maskinfabriken
Kjeldgaardsveg, PO BOX 143, DK-9300, Saeby, Denmark

Erling Foss Export
Thorsgade 59, DK-2200, Copenhagen, Denmark

United Milling Systems Ltd
Gamle Carlsbergsvej 8, DK-2500 Valby, Copenhagen, Denmark

Akron
531 04 Järpås, Sweden

Nirvana, AB Svegma
Box 400, 53500 Kvänum, Sweden

Ets Guy Moulis
Avenue de Castres, 81360, Montredon-Labessonie, France

Ets A.Gaubert
16700 Ruffee, 23 Rue Gambetta, France

Benson Et Cie
BP 14, 59550, Landrevies, France

S.A.M.A.P
BP 18, Horbourg-Wihr, 6800 Colmar, France

Argoud SECA
Le Mottier, 38260 La Cote St.Andre, France

Ets Champenois S.A
Chamouilley, 52170 Chevillon, France

Ets Claudien Beronjon
280 Rue de Alpes, 38290 Verpilliere, France

Societe Comia-FAO S.A
27 bd de Chateaubriand, 35500 Vitre, France

Goudard
77260 Le Ferte sons Jouard, France

Law Secemia
BP 15, 5 Rue de General de Gaulle, 60304 Senlis Cedel,
France

Moulis
80800 Montreden La Bressonie, France

Promill
BP 109, 28104 Dreux, France

S.A.M.A.P
1 Rue du Moulin, BP 1 Andolshein, 68600 Neufbrisach,
France

Ets Simon Freres
Rue Laurent Simon, BP 177, 501104 Cherbourg Cedex,
France

Tixier Freres
18120 Lurt Sur Arnon, France

Group de Recherche d'Echanges Technologique
213 Rue Lafayette, 75010 Paris, France

Campagnie Internationale pour le Development Rural
BP 1, Autreches, 60350 Cuise la Motte, France

Iruswerke Dusslingen
7401 Dusslingen, Postfach 128, Germany

Amos Machinenfabrik GmbH
Postfach 23, D-4930 Delmold, Germany

Ceccato Olindo Machine
Via Guistiniani 1, 35010 Arsego(SP), Italy

Ten Have Engineering bv
Industrieweg 11, Postbus 27, 7250, AA Vorden,
Netherlands

Buhler Pros Ltd
CH-9240, UZWIL 4, Switzerland

E.H Benthall & Co Ltd,
Maldon, Essex CM9 7NW, UK

John Gordon & Co Ltd
196 High Str., Epping, Essex, UK

Harrap Wilkinson Ltd
North Phoebe Str., Salford MS 4EA, UK

R.Hunt & Co Ltd
Colchester, Essex CO6 2EP, UK

Christy & Norris Co Ltd
Maldon, Essex CM9 7NW, UK

Colman & Co (Agr.) Ltd
Sudbury, Suffolk CO10 6NY, UK

Cornercroft (Agr.) Ltd
Conningsby, Lincs LN4 4SN, UK

Kamas Machinery Ltd
110 Hunslet Lane, Leeds, Yorkshire, UK

R.A Lister Ltd
Dursley, Glos GL11 4HS, UK

Miracle Mills
Franklin Road, London SE20 8JD, UK

Ramsomes, Sims & Co
Ipswich, Suffolk IP3 9QG, UK

Scotmec Ltd
PO BOX 31, Stockport, Cheshire SK3 0RT, UK

Turner Grain Handling Ltd
Benezet Str., Ipswich, Suffolk 1P1, 2JQ, UK

John Wilder (Eng.) Ltd
Hithercroft Works, Wallingford, Oxon OX10 9AR, UK

Figure 1-2. Lars-Ove Jonsson (1992).

Figure 3. Francois, M. (1988); Du Grain à la Farine, pages 14, 15, 52.

Figure 4. Lars-Ove Jonsson (1992).

Figure 5. Natural Resources Institute (NRI).

Figure 6. Kent, N.L. *Technology of Cereals,* (1983) page 19.

Figure 7. FAO Food and Nutrition Paper 47/1, (1989) page 85.

Figure 8. FAO Food and Nutrition Paper 47/1, (1989) page 6.

Figure 8b. NRI.

Figure 9. FAO Food and Nutrition Paper 47/1, (1989) page 66.

Figure 10. Bassey, M.W. and Schmidt, O.G. *Abrasive-disk dehullers In Africa; from research to dissemination,* (1989) page 128.

Figure 11. Borasio, L. and Garibaldi, F. *Illustrated glossary of rice processing machines,* FAO, (1957) page 32.

Figure 12. Manufacturer's instructions, CeCoCo.

Figure 13. Borasio, L. and Garibaldi, F. *Illustrated glossary of rice processing machines,* FAO, (1957) page 24.

Figure 14. Borasio, L. and Garibaldi, F. *Illustrated glossary of rice processing machines,* FAO, (1957) page 32.

Figure 15. Wimberley, J.E. *Paddy rice post-harvest industry in developing countries,* IRRI, (1983) page 138.

Figure 16. Garibaldi, F. *Rice milling equipment operation and maintenance.* Agr. Service Bulletin 22, FAO, (1974) page 68.

Figure 17. Wimberley, J.E. *Paddy rice post-harvest industry in developing countries,* IRRI, (1983) page 136.

Figure 18. Singer, C. (Ed.) *History of technology,* Vol 2, (1954) page 109.

Figure 19. *Small-scale maize milling,* Technical Memorandum No 7. ILO, (1984) Front cover.

Figure 20. Pinson, C. *A pedal-operated grain mill,* Rural Technology Guide 5, NRI, (1978) page 27.

Figure 21. Boyd, J. *Tools for agriculture: a buyer's guide to low cost implements,* ITDG, (1976) page 128. *Small scale maize milling,* Technical Memorandum No 7. ILO, (1984) page 79.

Figure 22. Boyd, J. *Tools for agriculture a buyer's guide to low-cost implements,* ITDG, (1976) page 128. Small scale maize milling, Technical Memorandum No 7. ILO, (1984) page 72.

Figure 23. Boyd, J. *Tools for agriculture: a buyer's guide to low cost implements,* ITDG, (1976) page 128. *Small scale maize milling,* Technical Memorandum No 7. ILO, (1984) page 81. Swedish Institute of Agricultural Engineering No. 359 and 418.

Figure 23b. SKIOLD Maskinfabrikken A/S.

Figure 24. *Cereal Processing,* Food Cycle Technology Source Book No. 3, UNIFEM, (1988) page 24.

Figure 25. Boie, W. *Introduction of animal-powered cereal mills,* GATE, (1989) page 15.

Figure 26. Ruston Service Manual, IRAQ, (1968) front page.

Figure 27. Lister Service Manual, (1987) front page.

Figure 28. Berglund, N. and Andersson, Y. Maskinlära, (1957) page 93.

Figure 29. NRI.

Figure 30. Fraenkel, P. *et al, Micro-Hydro Power: A guide for development workers,* IT, (1991) page 83. *Renewable energy for development,* SEI, Vol 14, No 2, Nov 1991, front page.

Figure 31. Carruthers, I. and Rodriguez, M.: *Tools for agriculture; A guide to appropriate equipment for smallholder farmers,* IT Publications, (1992) page 129.

Figure 32-35. NRI.

Figure 36-37. Lars-Ove Jonsson.

Figure 38. Nordenborg, M.O. et al, *Allmän arbetslära för jordbruket,* (1963) page 147. Nojonen, M-A. Lantbrukets arbetsmiljö, (1989) page 119.

Figure 39. Nojonen, M-A. *Lantbrukets arbetsmiljö,* (1989) page 66.

Case 3.1-3.20. 'Hammer-mill Operator's Manual' (Draft), Village Industrial Service, Lusaka, Zambia, (1992), and Lars-Ove Jonsson.

Illustration 1. *The rural solar mill,* 11th European Photovoltaic Solar Energy Conference and Exhibition (1992).

ALSO AVAILABLE FROM INTERMEDIATE TECHNOLOGY PUBLICATIONS

Small-scale Food Processing
A guide to appropriate equipment
Edited and introduced by Peter Fellows and Ann Hampton

This book provides information on the major food-processing technologies, divided by food group, including sugar confectionery, milk, meat and cereal-based products. The book also catalogues the necessary equipment, manufacturers and product details, and prices.

300pp   Paperback.   1992.   ISBN 1 85339 108 5.

Food Cycle Technology Source Books

Designed for people who have no technical background or previous knowledge of the technologies, the titles in this series offer information on existing ways of improving the technology of food processing and increasing the quality and range of foodstuffs produced. While not providing instructions for actual processing, these source books are intended to increase awareness of the range of technological options and sources of expertise, indicate the complex nature of designing and successfully implementing technology projects and diffusion programmes, and provide material for those training in this area. The series has been prepared by the *United Nations Development Fund for Women (UNIFEM)* and *Intermediate Technology* in recognition of women's special roles in these processes.

Oil Processing
UNIFEM
48pp.   Paperback.   1993.   ISBN 1 85339 134 4.

Fruit and Vegetable Processing
UNIFEM
72pp.   Paperback.   1993.   ISBN 1 85339 135 2.

Fish Processing
UNIFEM
80pp.   Paperback.   1993.   ISBN 1 85339 137 9.

Cereal Processing
UNIFEM
72pp.   Paperback.   1994.   ISBN 1 85339 136 0.

Root Crop Processing
UNIFEM
76pp.   Paperback.   1993.   ISBN 1 85339 138 7.